T0305809

Saffron (*Crocus sativus*)
Production and Processing

Editors

M. Kafi
A. Koocheki
M.H. Rashed
M. Nassiri

Center of Excellence for Special Crops (CESC)
Faculty of Agriculture
Ferdowsi University of Mashhad
Iran

CRC Press
Taylor & Francis Group
Boca Raton London New York

CRC Press is an imprint of the
Taylor & Francis Group, an informa business

First published 2006 by Science Publishers

Published 2010 by CRC Press
Taylor & Francis Group
6000 Broken Sound Parkway NW, Suite 300
Boca Raton, FL 33487-2742

ISBN-13: 978-1-57808-427-2 (hbk)

Visit the Taylor & Francis Web site at
http://www.taylorandfrancis.com

and the CRC Press Web site at
http://www.crcpress.com

Preface

The history of agriculture bears testimony to the great endeavors of Iranian farmers in domestication of many plants and animals. These farmers have long been farming in an environment with scarcity of resources, particularly water. Consequently, their practices were based on optimum water utilization through appropriate cropping patterns and enhancement of water use efficiency. Utilization of underground water through a unique system of water extraction called *qanat* or underground conduits is documented as a technology developed by Iranians in dry areas. Development of cropping systems based on low water requiring plants such as saffron, cumin and barberry is another example of indigenous knowledge of farmers in these areas. However, with the advent of new farm technologies in the last century based on genetic improvement, water availability and high input of chemical fertilizers and pesticides, indigenous technologies were replaced and the focus shifted to new crops such as maize, soybean, potato, etc. Local crop species were not able to compete economically with such high yielding species and hence they were neglected. Most of these neglected crops are in developing countries and not much support has been given by the international research organizations in terms of funding. This was also the case for crops such as saffron, cumin, medicinal plants etc in Iran.

The Center of Excellence for Special Crops was established in Department of Agronomy, Faculty of Agriculture, Ferdowsi

University of Mashhad with a mandate to conduct research on such crops. One of the first attempts of the center was to collect and document all the local and international literatures on saffron, cumin and barberry—the main crops of Southern Khorasan. The present book is one of these series which is based mainly on the results of research conducted locally. These findings have been analyzed comprehensively and compared with results of international origin. This book contains eleven chapters. In the first chapter, history, importance, acreage, production and utilization of saffron have been reviewed in general. In the second chapter botany and plant characteristics are discussed and in the third chapter ecophysiological criteria of saffron is analyzed. Production technology is discussed in the fourth chapter and water requirement is the topic of chapter five. Pests and diseases of saffron are reviewed in the sixth chapter and genetic and breeding aspects are the topic of chapter seven. Economic aspects of saffron is the main area of chapter eight and in chapter nine role of indigenous knowledge in crop production with an emphasis on saffron is discussed. In chapter ten processing, standards and chemical composition are reviewed and finally research strategies are the main topic of chapter eleven.

The Editors would like to place on record their appreciation to the Center of Excellence for Special Crops, Ferdowsi University of Mashhad for providing financial support which enabled preparation of this book.

M. Kafi
A. Koocheki
M.H. Rashed
M. Nassiri

Contents

List of Contributors

Alizadeh A.
College of Agriculture, Ferdowsi University of Mashhad, Iran

Bagheri A.
College of Agriculture, Ferdowsi University of Mashhad, Iran

Hemmati Kakhki A.
Iran Scientific and Industrial Research Institution, Khorasan Center

Kafi M.
College of Agriculture, Ferdowsi University of Mashhad, Iran

Karbasi A.
College of Agriculture, Zabol University

Koocheki A.
Center of Excellence for Special Crops, Faculty of Agriculture, Ferdowsi University of Mashhad, Iran

Molafilabi A.
Iranian Scientific and Industrial Research Organization, Khorasan Center

Rahimi H.
Khorasan Agricultural Jahad Research Center

Rashed-Mohassel M.H.
College of Agriculture, Ferdowsi University of Mashhad

Shahrokhi M.B.
  Khorasan Agricultural Jahad Research Center,
Vesal S.R.
  College of Agriculture, Ferdowsi University of Mashhad, Iran

CHAPTER 1

Historical Background, Economy, Acreage, Production, Yield and Uses

M. Kafi[1], A. Hemmati Kakhki[2] and A. Karbasi[3]
[1]College of Agriculture, Ferdowsi University of Mashhad, Iran
[2]Iran Scientific and Industrial Research Institution, Khorasan Center
[3]College of Agriculture, Zabol University

1-1 INTRODUCTION

Saffron (*Crocus sativus*) is the most valuable crop species in the world and is the only plant whose product is sold by gram. Saffron belongs to *Irridaceae* family and is geographically distributed in Mediterranean climates, East Asia (latitudes 30-50° N and longitudes 10° E to 80° W), as well as Irano-Touranian regions with low annual rainfall, cold winters and hot summers. While some species are grown for their beautiful ornamental flowers, cultivated species of saffron have great economic importance.

Saffron has unique characteristics, its flowers appear before any vegetative development, its growth starts in autumn and ends in spring, its seeds are sterile despite production of many flowers, and its flowers should be harvested early morning before sunrise (5, 18). Therefore, its production technology is more complicated compared to other crops and is mainly based on farmers' indigenous knowledge.

In this chapter the history of saffron, its economic importance, cultivated area and production as well as its uses will be discussed.

1-2 HISTORY

There are different theories about the origin of saffron. However, based on documented evidence it originated in Iran, most probably in Zagross and Alvand mountains.

Wild saffron, known as "*Gouishi*" or "*Kerkomise*" has similarities with domesticated species. For example corm, leaves, stamen, and style in *Crocus* is quite similar to Gouishi. However, because of its short style and low odor *Gouishi* has no economic value (1).

The oldest evidence about utilization of saffron dates back to "Achaemenian", an ancient Persian dynasty. They were engraved a list of different spices used in the court kitchen on a metal column which was erected in front of the Kings' Palace. Polien, the Greek historian at 2 BC., has recorded all the spices from this column. According to Polien, as much as 1 kg per day of saffron was used in the court kitchen.

In post-Islam literature and in Iranian dictionaries and books, the Arabic word "Saffron" has been used both as the name of a plant and its product. In fact the Persian word "Kerkom" was never used again in public conversations.

1-3 ECONOMIC IMPORTANCE

Saffron as the world's most valuable industrial/medicinal product is an important export commodity and is of great significance in Iran's agricultural economy. At present Iran with production of 65% of the world saffron is ranked as the largest producer. In 2002 the net income from exporting 121 tons of saffron exceeded 51 million US $ (13).

In Iran the importance of saffron besides its export value is related to high water productivity, rural employment, and high net profit compared to other crops (2). Socio-economic studies have shown that on average, production of 1 ha saffron in Iran (including all practices from sowing to harvest) is labor intensive and requires 270 person-days. Given the total production area of 47208 ha, it means 12.7 million person-days annually. Since 200 days employment is considered as a permanent job, saffron production in Iran provides up to 65,000 job opportunities per year and therefore, is considered as an important factor against migration from rural areas (2; 10).

1-4   ACREAGE, TOTAL PRODUCTION AND YIELD

At present saffron production is limited to Iran and countries with older civilizations such as Spain, Italy, India or Greece. Cultivated area, annual production and yield of saffron in the main saffron producing countries are presented in Table 1-1 (4). Data in Table 1-1 shows Iran's dominant position both in cultivated area and production among the main producing countries (9). In the following pages production trends in Iran will be discussed in detail as the global production of saffron hinges largely on the output of this country.

Table 1-1   *Cultivated area, production and yield of saffron in different countries for year 1999*

Country	Yield (kg ha^{-1})	Annual production (tons)	Cultivated area (ha)
Iran*	3.4	160.0	47208
Spain	6.5	29.2	4184
India	2.0	4.8	2440
Greece	5.0	4.3	860
Azerbaijan	4.3	3.7	675
Morocco	2.0	1.0	500
Italy	8.4	0.3	29

*In Duke (4) cultivated area and production in Iran was reported as 6000 ha and 30 tons, respectively, which was corrected based on national statistics.

Figure 1-1 shows that total cultivated areas of saffron in Iran increased by up to 3.9 folds during the last 12 years with mean annual growth rate of 22.4% (17). A similar trend was observed for total saffron production (Figure 1-2). However, total saffron production during the same period increased 2.8 times with a mean annual growth rate of 13.7% (17). Annual fluctuation in production (Fig. 1-2) compared to cultivated area (Fig. 1-1) is much higher which is mainly related to yield variation.

During the period 1990-2002 the highest and the lowest yield were 6.3 and 2.9 kg ha^{-1}, respectively, and shows high temporal variation (Fig. 1-3). Occurrence of drought is the main source of variation in saffron yield (14; 15). However, yield also depends on the age of saffron fields and is too low in the first year of establishment (17).

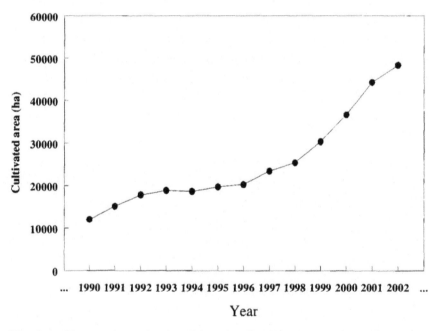

Fig. 1-1 Changes in total cultivated areas of saffron in Iran during the period 1990-2002.

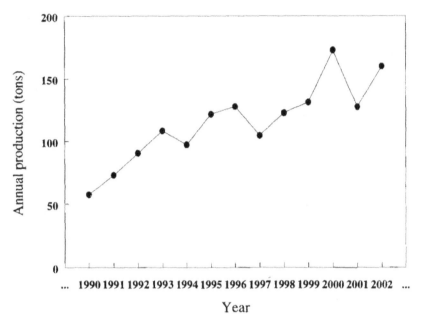

Fig. 1-2 Changes in total production of saffron in Iran during the period 1990-2002.

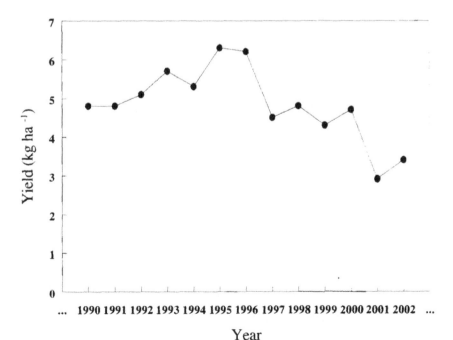

Fig. 1-3 Variation in saffron yield in Iran during the period 1990-2002.

While during the last few years saffron production has been extended to several parts of the country, Khorasan province is known worldwide as the most important saffron production area of Iran (9). Southern and central parts of this province have unique climatic conditions for saffron growth and therefore, 92% of the total country level production and 98% of cultivated areas of Iran belong to Khorasan (13).

1-5   USES

The three-branch style of *Crocus sativus* flowers is the most important economic part of the plant and known as saffron. This part of the flower contains fats, minerals and mucilage. Saffron odor is related to a colorless terpen essential oil as well as an oxigenous compound of sineole. The bitter taste of saffron is due to picrocrocin and picrocrozioide which are soluble in water and alcohol and hardly in chloroform. The origin of saffron color is crocin and produces glucose and crocetin ($C_{20} H_{24} O_4$) after hydrolysis.

While saffron is a well known spice, it has many other uses in food, pharmaceutical, cosmetics, and perfume industries and in production of textile dyes.

1-5-1 Food Industries

Saffron has a long history of use as a spice and for its wonderful color, odor and taste. During the recent years public interest towards using natural additives instead of synthetic chemicals has led to a breakthrough in using saffron as a natural flavoring in food industries (19). Today saffron is widely used in confectionary, alcoholic and non-alcoholic beverages and its use as a coloring agent for sausages, oleomargarines and shortening is allowed in USA. Saffron is also used in dairy products such as butter, cheese and ice cream for its color and for flavor improvement (8). Saffron should be used in specific doses depending on the food type; examples are given in Table 1-2.

Table 1-2   *Dose of saffron used in different foods (adopted from 16)*

Usage in foods	Dose (ppm)
Confectionary products	10
Non-alcoholic beverages	1.3-7.5
Alcoholic beverages	200
Ice cream	1.3-9
Meat	260
As a spice	50

1-5-2   Medicinal Uses

In traditional medication saffron has several properties such as relaxant, expectorant, exhilarating agent, digestion stimulant, spasm calmative, menstruation and fetus abortion. Saffron was also used against bloody diarrhea, fever, measles, hepatitis, liver and spleen syrose, urine infections, cholera, diabetes, and dermal diseases. In English pharmaceutical codex saffron syrup, saffron glycerin and saffron tincture are discussed (8).

Saffron is appetitive and facilitates digestion. Its essential oil is relaxant and could be useful in insomnia of nervous origin. Saffron for its effects on bronchus is used in chronic bronchitis and lung diseases. In India saffron is widely used for kidney, liver, vesica diseases and for medication of cholera (3; 20). External application of saffron tincture is useful for dermal disease such as impetigo (3).

Traditional knowledge of medicinal properties of saffron attracted scientific interest towards this spice and during the last decade several medicine research centers are investigating the biological and medicinal potential of saffron (12; Table 1-3). Reduction of blood bilirubin level and decrease in blood cholesterol and triglycierides after using crocin and crocetin are examples of new findings about saffron properties (11). Recently anti-cancer effects of saffron has been widely reported by many researchers (4).

Use of petals as edible color: During the recent years using synthetic coloring agents has been forbidden in food industries. Therefore,

Table 1-3   *List of some scientific institutes working on biological and medicinal aspects of saffron during the last decade*

Institute	Place
Amalar Cancer Research Center	Kerala, India
Central Institute for Food Technology	Mysore, India
Minnesota Medicine School	Minnesota, USA
Athens Agricultural University	Athens, Greece
Azerbaijan Academy of Sciences	Bakoo, Azerbaijan
Kyosho University	Foukata, Japan
Tokyo University	Tokyo, Japan
Morcia University	Morcia, Spain
National Institute of Health Sciences	Sparaky, Japan
Cortarro University	Cortarro, Mexico

identification and extraction of natural edible colors particularly those with plant origin are practised extensively.

After separation of colorful styles the remaining parts of harvested flowers of saffron are usually considered as waste with no special use. Recent studies have shown that these materials contain considerable amounts of antocyanine, a red plant pigment. This pigment together with other felavenoides of cell sap provides beautiful violet color of saffron petioles. After extraction this natural dye will change to red color in acidic conditions. Production of large amounts of saffron petioles after harvesting could be considered as a potential source of edible food color with antocyanine bases which is a proper substitute for synthetic red coloring agents (6; 7).

Methods of extraction of antocyaninie from saffron petioles and its use as an edible color were studied by Hemmati et al. (12). Their results showed that annually up to 10 thousands tons of petioles, which have a high content of extractable color as compared to other plant sources, are produced in the saffron production areas of Khorasan province of Iran. Evaluation of four different methods for extraction of color from saffron petioles showed that diluted HCl was the most proper solvent with the highest extraction efficiency and the lowest impurity in extracted product. Based on the results of this research extracted color could be used as a red edible color. However, since antocyanines are unstable compounds especially in response to pH variation, their storage condition is of great importance. Because of its sensitivity to pH antocyanines can only be used in acidic foods with pH lower than 4 (6; 7).

1-5-3   Use of Leaves as Animal Feed

One hectare of saffron produces about 1.5 tons of leaf dry matter. Where cultivated areas are high (e.g. 47 000 ha in Iran) annually considerable amount of leaf dry matter will be produced from saffron fields which could be used as dry forage. However, information about the feed value of saffron leaves is scarce. The possibility of using this forage source for native animals (sheep and goats) in Khorasan province was studied by Valizadeh et al (19). Their results showed that generally saffron forage had an intermediate quality and digestibility compared to conventional animal feeds such as alfalfa. The main deficiencies of saffron leaf dry matter were low digestibility due to high fibrous tissues and low protein and mineral content. However, low quality of this forage could be overcome by adding protein and energetic complements such as urea fertilizer (about 2% of leaf dry matter) or sugar beet molasses (19). While the cheap price of saffron forage and possibility of enhancement of its quality makes it a potential animal feed, more detailed studies about these aspects are required.

1-6   SAFFRON TOXICITY

At a high dose, saffron has narcotic and ecstasy effect and excessive delight which will finally lead to temporary paralysis. Abortion at overdosing with high risk of maternal death is reported (20; 3).

1-7   SUMMARY

Cultivation of saffron dates back to 2500 years ago. While this species is native in Greece and Mediterranean regions, it is believed that it originated in Zagros mountains of Iran. Cutivation of saffron was started in the central parts of Iran. However, at present main saffron production areas are located in Khorasan, a province in the south east of the country. Iran is the worlds' first saffron producer with up to 65% of the worlds' total production. Under optimum climatic conditions for growth and development saffron with its low water requirements, growth during winter and spring and high net income has an advantage over many crops. Saffron production is labor intensive and in highly populated rural areas provides permanent employments for many people. Saffron yield is sensitive to precipitation and shows high temporal variation in response to drought. While saffron is mainly known as a spice, its medicinal properties, especially its anti-cancer effects have been studied extensively during recent years.

References

1. Abrishami, M.H. 1997. Iranian saffron: historic, cultural and agronomic prospects: Astan Ghods Razavi Publishing Co. (Persian).
2. Agricultural Research, Education and Extension Organization of Iran, 1997. Evaluation of economic and technical aspect of saffron production in Iran.
3. Baker, D. and M. Neghi, 1983. Uses of saffron. Eco. Bol. 37(2): 228-236.
4. Duke, J.A. 1987. Handbook of medicinal herbs. CRC Press Inc. pp: 148-149.
5. Habibi, M.B. and A.R. Bagheri, 1989. Saffron, agronomy, processing, chemistry and standards. Scientific and Industrial Research Organization of Iran, Khoran Center.
6. Hemmati Kakhki, A. and S. Rahimi, 1994. Identification and extraction of antocyanine from saffron petals and evaluation of its stability in a model beverage. Project Report, Scientific and Industrial Research Organization of Iran and Iranian Institute for Nutrition and Food Industries.
7. Hemmati Kakhki, A. 2001. Optimization of factors affecting production of edible colors from saffron petals. Journal of Agricultural Science and Technology, 15: 13-21.

8. Hosseini M. et al. 2000. Evaluation of socio-economic consequences of saffron researches during the past 10 years. Scientific and Industrial Research Organization of Iran, Khoran Center.

9. Khorasan Jehad e Keshvarzi Organization, 2001. Statistics and Information Center.

10. Mohammadi, F., 1997. Economic evaluation of production and export situation of saffron and cumin. Agricultural Economics and Development Congress.

11. Nair, S.C., B. Pannikar and K.R. Panikkar, 1991. Antitumor activity of saffron (*Crocus sativus*). Nutrition and Cancer, 16: 167-172.

12. Neghi, M. 1999. Saffron (*Crocus sativus* L.). Harwood Academic Publishers, Amsterdam, 154 pp.

13. Sabzevari, 1996. Saffron, the red gold of desert. Agricultural Bank, No. 46.

14. Sadeghi, B. 1997. Effects of storage and corms planting date of flowering of saffron. Scientific and Industrial Research Organization of Iran, Khorasan Center.

15. Sadeghi, B. 1998. Effects of summer irrigation on saffron yield. Scientific and Industrial Research Organization of Iran, Khorasan Center.

16. Standard and Industrial Research Institute of Iran, 2000. Instruction for harvest and processing of saffron before packaging. Iran National Standard No. 5230.

17. Statistics and Information Center of Ministry of Jehad e Keshavarzi. 2002. Saffron production in Iran. Report No. 12.

18. Rashed Mohasel, M,H, A.R. Bagheri, M. Sadeghi and A. Hemmati 1989. Report of mission for Spanish saffron. Scientific and Industrial Research Organization of Iran, Khoran Center.

19. Valizadeh, R. 1988. Using saffron leaves for animal feeding. Project Report, Scientific and Industrial Research Organization of Iran, Khorasan Center.

20. Zargari, A., 1993. Medicinal plants (Vol. 4). Tehran University Publication.

CHAPTER 2

Saffron Botany

M.H. Rashed-Mohassel
College of Agriculture, Ferdowsi University of Mashhad

2-1 INTRODUCTION

Although, separation of plant species from each other has been demonstrated by different authors to a large extent, obscure points exist about the origin of saffron. It cannot be concluded precisely whether saffron, which has an old history to plantation ranica may perhaps have originated from wild species of saffron growing in South Africa and the Mediterranean basin. Ambiguity also exists about the Iridaceous phylogeny to which saffron belongs (7, 14, 16).

The following description presented by Judd et al. (1999), illustrates the formation and dispersal of new species. Any population of a plant species is able to produce plants with new heritable characteristics as indicated in figure 2-1. Hypothetically, over time, mutation may split this population in two. The resultant species establish their respective ancestor-progeny characteristics, while retaining certain similarities with each other. This is evident in the process of evolution. If later the population shown in figure 2-1, mutates in two Meta populations, one of which (right) produces red flowers and the other (left) produces woody stems. Red flowers and woody stems are evidence of two new and independent populations based on a simple new characteristic.

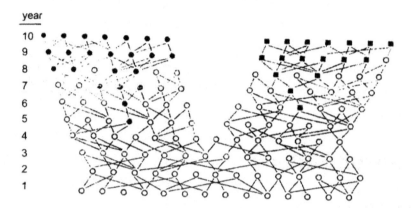

Fig. 2-1 Evolution of two hypothetical species. A mutation in progeny (left) results in production of woody species from herbaceous species and transmitted through generations. A similar mutation (right) produces plants with red petals (12)

o– white petals, herbaceous stem, smooth leaves, dry fruit, smooth testa, five stamens.

●– white petals, woody stem, smooth leaves, dry fruit, smooth testa, five stamens.

■– red petals, herbaceous stem, smooth leaves, dry fruit, smooth testa, five stamens.

This case may repeat within populations and in later populations new characteristics may be established (figure 2-2). For instance, some woody plants may have fleshy fruits while others possess spiny seeds–, some red flowers may have four stamens while others may have pubescent leaves.

Generally, these characteristics are termed "character" or "character state". The color of the flower has two states "white" and "red" and the character of the stem has "woody" and "herbaceous" states. New characteristics illustrate the occurrence of mutation while older characteristics say nothing about this event. After a period of time, these new characteristics may become the ancestors of younger characteristics. A group with a common ancestor and number of progenies is called "monophyletic". Similarly, we may combine figures 2-1 and 2-2 into figure 2-3. Thus, the term advanced or primitive refers to the location of the plant, and the plants on the upper part of the tree are more advanced.

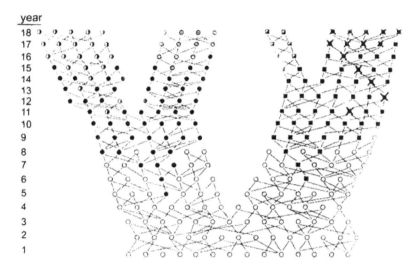

Fig. 2-2 The hypothetical plant of figure 2.1 after 8 more years in which 2 new outcomes resulted in new species (12).

◌–spiny seed coat ☺–woody stem, fleshy fruit ♥–pubescent leaves

◻–Four stamens ○–white petals ■–red petals ●–woody stem.

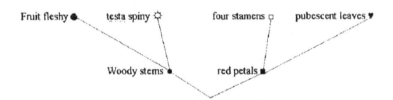

Fig. 2-3 Summary of figure 2-2 illustrates the events which resulted in the formation of new characters and species (12)

Cladistic analysis based on amino acids, ribulose diphosphate carboxylase in chloroplast confirmed the monophylic group of Iridaceae, but the phylogenetic of this family is not perfectly understood. Apparently, iridaceae is related to, but undoubtedly more advanced than liliaceae (2, 7, 12). In order to understand the genus of saffron and its position in the plant kingdom, we start from the iris family and then talk about different species of saffron in Iran and related boundaries, with particular attention to cultivated saffron which is a strategic crop in eastern Iran, and many other countries.

2-2 IRIDACEAE (SAFFRON FAMILY)

Terrestrial plants, herbaceous, perennials, rarely annuals (3 species of Sisyrhinchium). Underground stem rhizome, corm, or rarely bulb. Vascular bundles contain large polygonal crystals of calcium oxalate (not raphide) especially in Ixioideae which includs saffron (3, 24). A cluster of leaves may be observed in basal parts of the plants or leaves alternate, elongated, non pedicelate, with dorsiventral symmetry. Stomata anemocytic (9, 13). Inflorescence determinate, mainly scapose and severely modified scorpoid, sometimes reduced to a single flower. Flowers large and attractive, bisexual, bracteates, actinomorph or zygomorph (3, 6) 4 Petals 6, trimerous, free or united (17), sometimes outer perianth different from inners, aestivation valvate, or imbricate (4, 12). Stamens 3, alternate with inner perianth. Gynoecium mostly consist of 3 united carpels, 3 locule, stigma and style 3 branched, and sometimes petaloid with stamens in the abaxial side of each branch(3, 4, 12, 17). Few to numerous anatropous ovules in each carpel, axial placentation (4), occasionally unilocular with parietal placentation. Seeds nucleate type, endospermous, composed of hemicelluloses, fat, proteins, and mostly with no starch. Embryo enlarges but small with basal cotyledon. Fruit 3 locular capsule, seeds sometimes arillate or with fleshy testa (3).

There are 78 genera with 1750 species within this family. The main genera of the family are: *Gladiolus* (255 species), *Iris* (250 species), *Sisyrhinchium* (100 species), *Crocus* (85 species), Romulea (90 Species), *Geissorhiza* (80 species), *Babiana* (65 species), and *Hesperantha* (65 species) (12).

Iridaceae are subdivided into four tribes with the following characteristics (4, 12):

1. Isophysidoideae with superior ovary, e.g. *Isophys*.
2. Nivenioideae with separate blue flowers and some woody species, e.g. *Aristen*.
3. Iridoideae with nectar glands and long style divided below the anthers and extended into 3 folded stigmas. This tribe is also

subdivided into Sisyrhincheae in which the styles are alternated with stamens and Iris in which stigma and styles are petaloid.

4. Ixioideae such as Crocus, Gladiolus Romulea, and Geissorhiza which are monophyletic based on united perianth, having corms, flowers with no petioles, exine with porous sculptures, and closed sheaths. The ancestor of this tribe has nectar on its sepals (4, 12).

Similarities exist between Romulea, Syringodea, and Crocus. Romulea and Syringodea have ovaries below ground and actinomrph flowers with no petioles. Whether crocus cultivated mostly in Iran is closely related to Romulea and Syringodea is not clear. It is possible that convergent evolution resulted in such similarities (4).

The leaves of saffron are dorsiventral and navicular. Different species of saffron might be fall, winter, or spring flowering. The stigma and style of crocus varies considerably. The stigmata branches may sometimes be digitately divided. These branches are mostly yellow or red and rich in crocin and saffranol (5, 26). Cultivated saffron grows mostly in Iran, Spain, India and some time France, Italy and countries around the Mediterranean basin. Several hundreds flowers are needed to produce one gram of saffron. Saffron has medicinal value and some species of it are grown as ornamental (5). Cultivated saffron is of great value throughout the world. It is widely cultivated in Khorasan area in Iran. Most farmers in Khorasan are dependent on this crop. Some species of this genus also grow wild in Iran. In the following section we try to cover different species of Crocus sensus lato, and cultivated saffron sensus stricto.

2-3 *CROCUS* L.

An herbaceous plant that grows in the Mediterranean region and west Asia from 10°C W to 80°C E and 30°C N to 50°C S in Irano-Touranian areas with cold winters and low rainy hot summers, but rainy fall, winter, and spring. This genus is mostly distributed in the

Eurasian region, from Morocco and Portugal in the west and extends to Russia, Khyrgistan and ultimately Sin kiang in western China (15). Saffron grows actively from fall to late spring and survives in soil due to its fleshy hard corm, which is covered within coriaceous, membranous or reticulate sheath (6). Saffron is well adapted to the above conditions (16). This genus is composed of 85 species some with beautiful flowers are grown as house plants, in ornamental and home gardens, rock gardens and parks. Croci is dispersed in vast groups and produce beautiful scenery with different colors on the ground under favorable conditions (9). Saffron has a solid underground corm which terminates at adventitious roots (13). Leaves, bracts, bracteole, and flowering stalk are enclosed and protected inside a number of membranous sheaths called cataphylls which apparently originate from corm. In fact cataphylls and true leaves originate from actively growing buds and the former are around 12 to 14 to protect young leaves and flowering stalk (1). Leaves 5-30, 10-18 cm long, navicular, sword shaped, and graminaceous type. Leaves synanthus or hysteranthus (16). Flowers are solitary or few within cataphylls. Spathe subtending pedicel is present or absent, If spathe is present, it is usually located in the middle of the flowering stalk. Floral spathe (bract and bracteole) diphyllus or monophylus depending on the presence or absence of bracteole. Elongated funnel shape or cup shape flowers have mostly vivid colors such as white, golden, purple, blue and variegated (13, 20).

The Crocus species start their growth early in the fall rainy season by producing flowers/leaves. However, in some species flowering delays until winter or spring. Based on flowering period different species of Crocus are divided as follows (8):

1. Spring flowering species e.g. *Crocus biflorous*, *Crocus aureus*, *Crocus almehensis*.

2. Fall flowering species e.g. *Crocus sativus*, *Crocus caspicus*, *Crocus gilanicus*, *Crocus cancellatus*, *Crocus longiflorus*, *Crocus hausskenechtii*, *Crocus speciosus*, *Crocus cartwrightianus*.

3. Winter flowering species e. g. *Crocus michelsonii*, *Crocus korlkowii*.

Examples are mostly from the species present in Iran.

Different species of Crocus (exept *C. sativus* which is triploid and seedless) are primarily reproduced by seeds. Once they are established they reproduce and spread via corms. The corms divide after the leaves senesced. Crocus growth is sympodial and every year produces flowering stalks from apical or lateral corms (6). The relationship between leaf production, flowering, time of flowering, and also cytological information, corms covering, position of leaves, bracts, flowers and seeds result in dividing this genus into different subgenus (16).

Perianth consists of 6 tepals in two similar rows. Stamens are 3 and smaller than perianth, anthers longer than filament and elongated outward. Gynoecium's 3 locular each with several ovules, ovary inferior, stigma with 3 red or yellow arms usually broader at free ends, style long and yellow. Fruit a 3 septate capsule (16, 25). Floral formula is as follows:

Usually, the soft bracts are more or less transparent and do not enclose the perigone tube completely. Seed coat is covered by dense papilla. The chromosomes number in different species have reported $2n = 12, 14, 16, 26$ but in cultivated triploid saffron $3n = 24$ (16, 25). Figure 2-4 shows a diagram of a saffron flower and a different part of cultivated saffron.

Crocus is very similar to Colchicum but the latter has 6 stamens. The main center of croci distributions are in Turkey, Greece and Flora Iranica region (20). Since we plan to discuss cultivated saffron, its probable ancestor, and wild species of saffron in Iran and adjacent boundaries, we concentrate on the species belonging to these groups. We also come across different species of saffron after flowering stage or during flowering stage. In each case we have to identify the species. The following two keys adapted from other keys and tested can help us to identify the above mentioned species in Iran.

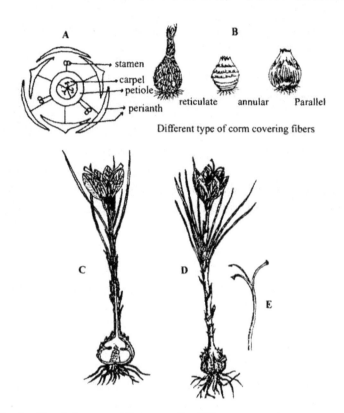

stamen
carpel
petiole
perianth

reticulate   annular   Parallel

Different type of corm covering fibers

Fig. 2-4 A—Floral diagram of Crocus. B—Different type of corm covering in *C. sativus*. C—Longitudinal section of C. sativus. D—Complete plant of *C. sativus*. E—Style and stigmatic branch of C. sativus.

I. Key for Crocus species based on vegetative growth and post floral structure (13, 20, 23). 1- Corm sheath membranous or leathery, without conspicuous fibers..2

　　1. Corm sheath rough or fine, with reticulate or parallel fibers...5

　　2. The width of mature leaves 4 to 5 mm................................3

　　2. The width of mature leaves 1 to 3 mm................................4

　　3. Leaves navicular with few distinct ribs either side of the keel...*C. Almehensis*

　　3. Leaves ± wide without ribs in either side of the keel...............
　　...*C. speciosus*

4. Capsule ± spherical, corm tunics is not quite annular at the base..*C. caspius*

4. Capsule cylindrical or elliptical, corm tunics is quite annular at the base...*C. biflorus*

5. Corm with 9-10 leaves..6

5. Corms with 3-7 leaves..7

6. Leaves grayish, no distinct vein either side of the leaves, corm tunic with a mat of reticulate fibrous sheath........................9

6. Leaves green, with one or more distinct veins either side of leaves, corm tunics with parallel or ± reticulate fibrous ....*C. korolkowii*

7. Corm tunics reticulate and rough......................*C. cancellatus*

7. Corm tunics fine ± reticulate or parallel................................8

8. Corm tunics a thick mat ± reticulate fibers..........*C.michelsonii*

8. Corm tunics with very fine and ± loose parallel fibers............ ....*C. gilanicus*

9. Corm 5 cm. diameter, capsule 1.5 -3 cm seed 3 -4 mm, diploid...10

9. Corm 4 cm or less, capsule small and elongated, seedless or rarely with seeds, triploid....................................*C. sativus*

10. Corm 1–2 cm in diameter............................*C. cartwrightianus*

10. Corm 3 cm in diameter................................*C. hausskenechtii*

II. Key for Crocus species based on floral structure (13, 20, 23).

1. Flowers yellow, sometimes outer portion bronze....................2

1. Flowers other colors, sometimes yellow at throat..................3

2. Leaves (7-)10 – 20, 1.5 – 3 mm width, corm tunics fibrous.... ....*C. korolkowii*

2. Leaves 3-4, 4-5 mm width, corm tunics membranous...... ....*C. almehensis*

3. Flowers and leaves produce simultaneously in spring............4

3. Flowering fall or early winter, corm tunics at leaf stage fibrous or membranous without ring at the base................................5

4. Corm tunics with a mat of fine fibers, style whitish.........
....*C. michelsonii*

4. Corm tunics membranous or coriaceous with distinct rings at the base, style usually yellow or orange red...............*C. biflorus*

5. Style 3 lobate or somehow lobate..6

5. Style with 3 arms...8

6. Flowering after leafing, flower throat dark yellow.......*C. caspius*

6. Flowering before leafing, flower throat not yellow..................7

7. Style barely or with indistinctive lobes, creamy or yellow, corm with fine tunics, ± fibrous.....................................*C. gilanicus*

7. Style with 3 distinctive lobes, yellow or orange red, corm with thick tunics and fibrous...9

8. Corm tunics with thick reticulate fibers leaves grayish...*C. cancellatus*

8. Corm tunics membranous or coriaceous, annular at base, leaves dark green...*C. speciosus*

9. Flowers white or bluish, inner and outer perianth unequal style red to orange, style arms 3 to 15 mm...........*C. hassknechtii*

9. Flowers lilac, purple or bluish white, inner and outer perianth equal, style 10 to 35 mm..10

10. Flowers blue violet or pink, stigmatic arms broad, 25 to 35 mm triploid (3n = 24)...*C. sativus*

10. Flowers pale, lilac to whitish color, stigma arms 27 mm, diploid (2n = 16)...*C. cartwrightianus*

2-4   CULTIVATED SAFFRON (*CROCUS SATIVUS* L.)

Syn. (16). *C. sativus var. officinalis* L.

C. officinalis var. *sativus* Huds.

C. autumnalis Smith

C. sativus var.*cashmirianus* Royle

C. orsinii Parl

C. sativus var. *C. orsinii*(Parl) Maw (Ro)

Saffron is derived from the Hebrew word "sahafaran" meaning thread. The Arabic word zafaran means yellow. Assumingly these words and two Greek words croci and karkom (yellowish), also krokos and zaffer are the roots of todays saffron and cultivated crocus (11).

Saffron is 20-30 cm tall. Corm spherical and compact, flat at the base, about 5 cm diameter, corm tunics with fine reticulate fibers extended upward about 5 cm above the neck of the plant (20, 23, 25). Cataphylls 5 to 11, white and membranous and protect newly grown leaves (15, 16). True leaves 5 to 11, synanthus or hysteranthus (some people believes if we start irrigation early leaves appear before flowering), Leaf straight and green smooth or ± pubescent, 1 to 3 mm diameter. Fragrant flowers about 1 to 4 from each corm, flowering during November to December, dark lilac or violet to purple rarely white with spots or reticulate venation at collar (16, 19), throat white to lilac and pubescent. Cataphylls, bracts, and bracteole are present, white and membranous, unequal, and tapering toward the end. Hairs are present where filament adnate to perigone tube. Perigone tube 4 to 5 and sometimes 12 cm long, perianth segments are almost equal, 3.5 to 5 cm length and 1 to 2 cm width (20), oblanceolate to ovate. Filament 7 to 11 mm white, yellow, or purple, anther 15 to 20 mm elongated, and yellow. Style is divided into 3 vivid red stigmatic broad and club shaped arms, each arm is about 25 to 35 mm long, and originates from upper half of anther or at least half portion of perianth segments. Ovary with 3 fused carpel, adjacent or below the ground level (see figure 2-4). The floral formula in this species is as follows.

$$+ \; \female \; P_{(3+3)} \; A_3 \; \overline{G}_{(3)} : \text{Capsule}$$

Capsule and seeds do not produce in it, or rarely we may see it. Cultivated saffron is triploid (3n = 24) and sterile (16).

In saffron, the development of root, stem, and leaf is from October till February. Producing of new corms starts after flowering in December. During May the leaves start yellowing and we may use it as forage for livestock. From May until September the reproductive organs differentiate. However, corms seem to be dormant.

Saffron has an ancient origin and is well known as a cultivated crop. Saffron is cultivated mostly in Iran, Spain and India. Cultivation of saffron in Iran dates back to before Christ. It is believed that this plant has been cultivated in Palestine during the prophet Solomon and during Jesus they took it from Jerusalem to England. Saffron is apparently endemic to Greece and Mediterranean regions, but its actual origin has been lost like several other ancient crops (18). Saffron has about 80 species distributed from Mediterranean region to China. Evidently saffron had more acreage in ancient Iran compared to present which is almost restricted to south Khorasan and portion of Fars in Estahbanat. However, presently saffron acreage is expanding considerably to other areas in Iran. In other countries such as, Turkey, Italy, India, Pakistan, France, and the United States saffron is being cultivated in a limited amount. In Iran about 95% of cultivated saffron is in Khorasan (mainly Torbatehaydarieh, Ghaen, Gonabad, Ferdows, and Birjand) area and another 5% is cultivated in Estahbanat in Fars, Esfahan and other parts of Iran (20, 21, 22, 23, 25).

There is less ambiguity concerning the origin of saffron from wild saffron in Greece (10). Based on plant international codes, the name of this species should change as it has been published before. However, for scientific purposes, it seems more rational that cultivated saffron which is cultivated in a wide area be considered a new species, and the familiar name with broad spectrum usage be applied to this plant with valuable commercial product. This practical method should not provide suspicions. We have similar other cultivated plants which presently differs considerably from their wild ancestors, Garlic (*Allium sativa*), onion (*Allium cepa*) and some cereal crops are such examples (16).

2-5 WILD SAFFRON *(CROCUS CARTWRGHTIANUS)*

Syn.(16) C. *sativus* Cibth & Smith

C. *sativus* Linn. Var. *cartwrightianus* (Herb) Maw.

2-6 JOE QUASEM SAFFRON (*CROCUS PALLASII* SUBSP. HAUSSKNECHTII)

C. pallasii subsp. Haussknechtii (Boiss & Reut. Ex Maw) B. Mathew
Syn. *C. sativus* var. haussknechtii Boiss & Reut. Ex Maw
C. hausskenechtii (Boiss & Reut. Ex Maw) Boiss.

Corm oval to compact spherical, 30 mm diameter, lower portion somehow flat, corm tunics fibrous, perfectly reticulate and extended about 10 cm upward and covered perigone tube. Cataphylls 3 to 5, white and membranous. Leaves 7 to 17, rarely 5 (17). Fall flowering, synanthus or leaves may appear shortly after flowering and may remain on the plant until next growing season, grayish green, 0.5-2 mm wide, sometimes pubescent at margin or keel (20, 25). Flowers fragrant, 1-6, pale lilac, pinkish lilac, or purplish blue, veins usually darker, perigone usually white, lilac, purple and pubescent, 4-10 cm. Spathe present. Bracts and bracteole also present, unequal, tapering ends, white membranous, located within cata, ':ylls, their length are about 4.5-6 cm. Perianth segments obovate, ± emarginated, round or blunt, rarely sharp pointed, 3.2 – 4.2 × 0.8 – 1.4 cm. Inner perianth usually looks smaller than outer perianth (16, 25). Filament 3 to 13 cm, white and glabrous or sparsely haired, Anther yellow 10 to 20 mm long. Stigmatic arms abruptly broad, vivid red at distal end. The point of dividing style located in upper half part of the anther. Stigma arms are 3-15 mm long and about the same level as perianth segments, broad at the free end and gradually slender toward the base, as shown in Fig. 2-6. Fruit an elliptical capsule, 15-30 mm long and 7-10 mm wide located on a short peduncle up to the ground level or few mm. higher than soil surface (16). Seeds are irregularly spherical, 3-4 mm in diameter, red to purplish and having a short caruncle (16, 25), the testa is covered with dense hairs, raphe and a short ridge along the seed (sometimes look like a wing) are also present. Diploid with $2n = 16$ chromosomes.

Flowering time is from October to November. We may find this species in open and rocky dry hills between short oak trees at 1300 – 2300m altitude (16). This saffron is mostly distributed in west of

Iran, northeastern portion of Iraq, and southern part of Jordan. In Iran this species has been observed between Hamadan to Kermanshahan, Arak, Shiraz and Kazeroun at 2000 -2300m altitude (25). This saffron is also one of the most similar to cultivated saffron. The type specimen of this plant has been identified in Kurdistan area in Iran (16). This species is also similar to C. *cancellatus* with its style divided into several yellow or orange branches (25).

2-7 ALMEH SAFFRON (*C. ALMEHENSIS*)

Corm tunics coriaceous, parallel and sharp pointed fibers at the apex, several distinct annulus at the baseare serrated on the upper edge. Height between 5 and 10 cm. Leaves 3-4, Synanthus, 3-4 mm width, grayish green with marginal hairs, conspicuous nerves may be seen either side of the keel. Spathe absent, bract and bracteole exserted from cataphylls 4, 4-6 cm long. Spring flowering, 1-3, Perigone tube 5-7.5 cm. Perianth segments almost equal, 2.3 – 3 × 0.6 - 0.9 cm elliptic to oblanceolate, glabrous, orange to orange yellow and bright (9, 13, 20). Anthers 0.8 -1.2 cm, yellow; filament 0.5- 0.7 cm, glabrous. Style yellow, same size as stamen, terminated into 3 red to orange stigmatic arms frilled at free end. Fruit capsule, 1.7-2.1 cm almost on the ground level at maturity. Seeds brown, 0.25 to 0.3 cm long (13, 25). This species has been collected from Almeh, Rabate-ghara-bil and national park of Gorgan at 1600-1900 m (21, 22). Time of flowering late winter after the snow melts, sometimes inside the snow (figure 2-7). This species is endemic around mentioned areas and for the first time it was identified in Iran in 1973. Diploid and 2n = 20 chromosomes (15, 23).

2-8 VIOLET SAFFRON (*C. MICHELSONII*)

About 10-15 cm tall. Corm relatively elongated, 2 to 2.5 cm in diameter, membranous tunic composed of several layers of coarse netted brown fibers encircled at the base of the stem. Leaves 4-7, 0.2 cm width, rarely ciliate on the margin of lamina and keel, synanthus,

This species is found in some areas in Iraq, and in Iran it was found in Gangale Golestan, Marand, Tabriz, Uromieh, and Kermanshahan at altitudes of 800-2700 m at rocky hillside and abandoned ranges with plant debris during February to May. This is also a diploid species and 2n= 20 chromosomes (20, 23, 25).

2-10 ZAGROS SAFFRON(*C. CANCELLATUS*)

The corm tunica includes high neck and reticulate fibers at free ends. Cataphylls 2-3, usually membranous and white. Leaves hysteranthus, sometimes synanthus, appear in winter or spring, 3-5, 0.1-0.25 sometimes up to 0.45cm width, grayish, smooth but somehow rough, occasionally pubescent on abaxial or edges of keel (20, 25), one or more vein may observe on either side of the keel. Spathe absent. Bract and bracteole 3-6 cm just exserted from cataphylls. Perianth segments equal, oblanceolate, 0.6-1 × 2.3-4.7 cm, glabrous, white to violet blue, nerves abaxial and throat of perigone dark purplish and on adaxial looks parallel white or yellow (figure 2-10). Anthers 1.3-2.1 cm long, yellow or white; filament 0.3-0.4 cm yellow or orange. Style the same size or slightly taller than stamens and usually (rarely not) divided into several yellow or orange stigmatic arms. Fruit cylindrical capsule, 2-3 cm long, located on a short peduncle above the ground level. Seeds 0.6 cm, elongated, marginated, and reddish brown.

This species is widely distributed across Europe and Anatolia region. In Iran it was found in northwest and northwestern parts in Arak, Khansar, Hamadan, Khoy, Uromieh and Kermanshahan mostly in side hills and rocky plains of Zagros area at 1500 to 2900 m altitude. It seems that this species is also cultivated in some areas in Iran (25).

Flowering time is from September to November. However, the leaves usually appear during winter and spring. Diploid species, 2n= 12 chromosomes (20, 23).

2-11 ZIBA SAFFRON (*C. SPECIOSUS*)

Corm is spherical and flat at the base; tunics membranous and thinner than other croci, with crossed gaps, 1 or 2 distinct annulus with entire or serrated edges may be observed at the base, some adventitious daughter corms may form at basal part of mother corm. The plant is 12 cm high (25). Leaves 3-4, hysteranthus, 0.4-0.5 cm width, may reach up to 40 cm long till the next growing season, pubescent at margins and abaxial keel. Spathe absent. Bract and bracteole 4-10 cm long, not exserted from cataphylls. Flowers solitary, perigone tube 6-15 cm, perianth segments ± equal, 1.1-1.7 × 3.7-6 cm, purplish blue, with reticulate veins orientations, the base of external perianth segments purple. Anther yellow, 1.5-2.4 cm, usually glabrous or having yellow or white hairs; filament 0.5-0.7 cm. Style keeled, at the same level of anther, terminated to 3 bright orange stigmatic arms, higher than anther (figure 2-11). Fruit a capsule, 2-2.5 cm at maturity and above the ground level. Seeds reddish brown, 0.3 cm long.

This specimen is found in areas of Azarbayejan, Gilan, Rostam abad, Tonekabon, Kalardasht, and gangale Golestan between 1000-2000m altitudes, among woody bushes, ranches, and grasslands. Flowering time is from September till November. This saffron is also diploid and 2n = 12 choromosomes (23).

2-12 CASPIAN SAFFRON (*C. CASPIUS*)

Corm ovate, tunics membranous, bright brown with longitudinal cracks, fibrous at base along with scattered radial fibers at the bottom (13). Cataphylls usually 3, leaves 2-7, glabrous, 1.5 mm wide, synanthus. Spathe absent (13). Bract and bracteole exserted from cataphylls, 4-9.3 cm, perianth white or slightly purple (15, 16, 20, 23, 25), segments almost equal, glabrous, 0.6-1.5 × 2-4 cm, obovate, outer perianth tapering distally; perigone tube 4-18 cm yellow to orange at throat. Anther yellow, 1.2-1.6 cm and longer than filament. Style keeled, clearly above anther, stigma arms 3, yellow

Fig. 2-10   Zagros saffron (*C. cancellatus*)

Fig. 2-11   Ziba saffron (*C. speciosus*)

Fig. 2-12   Caspian saffron (*C. caspius*)

glabrous and drooping. Fruit a capsule, 1-1.3 cm ± spherical, on the ground level or slightly below it at maturity. Seeds about 0.3 cm, dark brown, ± smooth but having angles (20, 23).

In Iran it was found in Gorgan, Amol, Haraz valley, Farahabad in Sari, and Bandare Anzali in grasslands and marginal parts of Caspian region forests at 100-1300 m altitudes (25). Flowering time September – November. Diploid, 2n = 24 chromosomes (23). Type specimen in Leningrad (see also figure 2-12).

2-13 GILAN SAFFRON (*C. GILANICUS*)

Corm almost spherical, flat at the base, about 1cm in diameter: tunics thin with parallel fibers. Spathe present, 0.9 cm. Cataphylls 3-4 and whitish. Only bracts within cataphylls, 0.8 cm long. Hysteranthus, Flowers solitary, perianth segments ± equal, flower segments oval, elongated, sharp pointed, 0.4-0.5 × 2.2-2.3 cm, nerves bluish purple; Perigone 3-4 cm, pubescent where stamens adnate perianth. Filaments and anthers equal, 0.5 cm. Style with 3 or 4 stigmatic branches at the same level of anther or slightly higher than anther. Capsule and seeds did not observe. Flowering time October. Type specimen was seen in 1973 between Khalkhal and Asalem in Azarbayejan at 2400m altitude (20, 25) (see also figures 2-13, and 2-14 for more details).

2-14 *C. KOROKOWII*

Corm elongated spherical; tunics membranous, fibers parallel and somehow interwoven. 10-20 cm tall. Cataphylls 3-5, white, membranous, brownish green at tips. Leaves 7-20, synanthus, 0.15 - 0.30 cm wide, one or more veins either side of the keel, leaves about the same level as perigone tube during flowering but clearly taller afterward (15, 16, 20, 23, 25). Spathe absent, bract and bracteole 4-10 cm and exserted fom cataphylls. Perigone tube 5-12 cm and perianth throat naked. Perianth segments yellowish orange, 0.6-1.3 × 2-4 cm, outer segments somehow taller and brownish on abaxial.

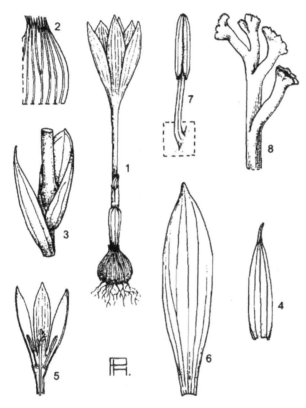

Fig. 2-13   Gilan saffron (*C. gilanicus*) –1: Habit × ½. 2: Corm tunic × 24 . 3: Spathe and bract × 6. 4: Spathe (flattened) × 6. 5: Flower × 1 ½. 6: Perigone segment × 4 ½ . 7: Stamen × 6. 8: Stigmas × 12

Anthers converged, orange, 0.9-1.3 cm long; filament 0.4-0.6 cm, glabrous or sparsely hairy. Style keeled, up to midsection of anther, equal or taller than stamens, yellow or orange with 3 upright orange stigmatic branches (16, 23). Fruit a capsule, 1.3 × 2.1 cm, cylindrical or elliptical, on the ground level or slightly within soil at maturity. Seeds trigonal, brownish red, 0.3 cm. This species is mostly distributed in Afghanistan, Pakistan, and Tajikestanin rangelands and rocky slopes. It was not found in Iran, but it is possible it existed in eastern or northeastern parts of Iran (20). Flowering time is in late February to March, Diploid, 2n = 20 chromosomes. Type specimen grown in Leningrad.

Fig. 2-14   *C. korokowii*

SUMMARY

In this chapter the botany of Iridaceae (including saffron genus), cultivated saffron, wild saffrons existed in Iran, and the possible ancestors of cultivated saffron have been studied and to some extend introduced. The historical evidence indicates that either in the past or present, Iran was the bed of cultivated saffron. Nowadays, the cultivation of saffron is almost specific to Khorasan. Recently, due to good saffron marketing, the acreage of saffron is increasing tremendously. Research on this strategic plant is important from different aspects including nutritional, medicinal, industrial, and economical purposes. It will also benefit the farmers in the region and for the sake of introducing foreign exchange into the country to work more seriously in obtaining more information about this valuable plant. Several wild type of saffron grow in Iran but all of them are diploid in spite of cultivated saffron which is triploid. It seems that the origin of this plant is from Greece where wild saffron

(*Crocus cartwrightianus*) exist. However, the saffron named *C. hausknechtii* which exists in western Iran also has similarities to cultivated saffron. Phyllogenetically, the relationship of other saffron that existed in Iran remains unknown.

We may conclude that the origin and knowledge of how cultivated saffron derived from its ancestor needs more investigation.

Reference

1. Ahuja, A., Skoul, and G. Ram. 1994. Somatic embryogenesis and regeneration of plantlets in saffron (*Crocus sativus* L.) Indian Journal of Experimental Biology, 52: 135-140.

2. Chase, M. W., M. R. Duvall, H. G. Hills, J. G. Conran, A. V. Cox, L. E. Equiarte, J. Hartwell, M. F. Fay, L. R. Claddik, K. M. Cameron, and S. Hoot. 1995. Molecular systematic of Liliaceae. In Monocotyledons: systematics and evolution, (ed.) 109-137. Royal Botanical Gardens, Kew.

3. Cronquist, A. 1981. An integrated system of classification of flowering plants. Columbia University Press. New York. Pp. 1211-1214.

4. Dahlgreen, R. M. T., H. T. Clifford, and P. F. Yeo. 1985. The families of the monocotyledons. Springer-Verlag, Berlin.

5. Duke, J. A. 1987. Hand book of medicinal herbs. CRC Press Inc. Pp. 148-149.

6. Ghahraman, A. 1995. Plant systematic, cormophytes of Iran. Volume 4: second printing, Iran University Press.

7. Goldblatt, P., 1990. Phylogeny and classification of Iridaceae. Annual Missouri Botanical Garden. 77: 607-627.

8. http://www.Web.obduedu/webroot/instr/sci/plant.nsf/payes/saffron

9. http://www.Botany.com/crocus.htm/

10. http://www.humorscope.com/herbs/saffron.htm/

11. http://www.suite/o/.com/article.cfm/5710/78682.

12. Judd, W. S., C. S. Campbell, E. A. Kellogg, and P. F. Stevens. 1999. Plant systematics. A phylogenic approach. Sinauer Associates Inc. Pp.191-192.

13. Komarov, V. L. 1968. Flora of the U.S.S.R. volume ²V: Printed in Jerusalem by IPST Press. Pp. 380-390.

14. Lewis, F. U. 1954. Some aspects of the morphology, phylogeny, and taxonomy of the South African Iridaceae. Annuals of the South African Museum. 40: 15-113.

15. Mathew, B. 1999. Botany, taxonomy and cytology of *Crocus sativus* L. and its allies. In saffron (*Crocus sativus*) (ed.) Harwood Academic Publisher, Pp. 19-30.

16. Mathew, B., and C. A. Brighton. 1977. Four central Asian species (Liliaceae). The Iranian Journal of Botany 1(2): 123-135.

17. Mozaffarian, V. 2001. Plant classification. Book 11, Morphology and taxonomy. Amir Kabir Publications. Tehran, Iran.

18. Papnicolaou, K., and E. Zacharof. 1980. Crocus in Greece, New taxa and chromosome numbers. Bot. Not. 133: 155-163.

19. Pigantti, S. 1982. Flora ditalia. Edagricola Press, Italy. Pp. 421-422.

20. Rashed-Mohassel, M. H. 1989. The identification and distribution of saffron genus in Iran. Proceedings of First Saffron Conference, Eslam Abad, Ghaen.

21. Rashed- Mohassel, M. H. 1993. Khorasan vegetations. Volume 1. Ferdowsi University of Mashhad Publication, Mashhad, Iran.

22. Rashed- Mohassel, M. H. 1994. Khorasan vegetations. Volume 2. Ferdowsi University of Mashhad Publication, Mashhad, Iran.

23. Rechinger, K. Flora Iranica. 1975. Iridaceae. Volume 112. Academische Druck-U. Verganstalt, Granz, Austria.

24. Rudall, P. 1994. Anatomy and systematic if Iridaceae. Botanical Journal of Linnean Society. 114-121.

25. Wendelbo, P. 1977. Tulips and irises of Iran and their relatives. Botanical Institute of Iran, Botanical Garden. Tehran, Iran.

26. Zargari, A. 1993. Medicinal plants. Volume 4, Tehran University Publications.

CHAPTER **3**

Saffron Ecophysiology

M. Kafi
College of Agriculture, Ferdowsi University of Mashhad, Iran

3-1 INTRODUCTION

Saffron is a plant with wide ecological, physiological and
phenological differences when compared with other conventional
cultivated plants. These differences could be classified as:

- Flowers appear before development of other plant organs, this
  process is dependent on the food reserves stored in the corms
  (28, 29).
- Beginning of flowering periods coincides with cold
  temperature in the fall, while in most other cultivated plants,
  this period starts with the warm and moist season (5, 35).
- Economic yield in most conventional plants are vegetative
  organs, seeds, roots or flowers as a whole, while in saffron only
  a small part of the flower – stigma, is economic yield (1, 30).
- For most of conventional cultivated plants, Harvest Index
  (HI) is calculated as a proportion of biological yield which is
  used as economic yield. This Index varies widely in crops such
  as cereals, pulses and oil crops ranging from 30 to 60% of
  biological yield, while HI in saffron is less than 0.5% (20).

- Response of conventional plants to cultivation practices and inputs, including water, during the growing season and particularly in the summer is a positive one, whereas the response of saffron to the same shows a specific pattern (31).

In this chapter findings related to ecophysiological criteria of saffron have been reviewed. However, there is still a great deal of controversy related to ecological, phenological and physiological attributes of this plant, which could be a subject of future investigations.

3-2 CLIMATIC REQUIREMENTS

3-2-1 Temperature

Although saffron grows well under temperate and dry climates, its vegetative growth coincides with cold weather and freezing conditions. Usually the maximum temperature for October, November and December in the southern parts of Khorasan—the main saffron growing area of the Iran does not exceed 20°C, while the minimum temperature reaches 0°C. Base temperature for saffron has not been recorded in literature, but Javanmard et al. (22) recorded the minimum temperature tolerated by this plant at -18°C. However, local populations of saffron in Torbate Hydarieh, the extreme northern area for saffron production in Khorasan tolerates temperature as low as -22°C. Normally the climatic conditions of Southern Khorasan in the areas of Ghaen, Gonabad, Ferdows and Birjand are regarded as the standard type of climate required by saffron. Climatic parameters such as temperature, rainfall and probability of freezing occurrance in the months of October and November is well documented (36). Since flower development mainly proceeds under soil conditions, soil temperature plays an important role in flower production. Planting depth is normally recommended to be in the deeper layers of the soil (15 cm). This may be associated with low variability of soil temperature at this depth

(30). In an experiment, it was noted that saffron was better adapted to flood irrigation than furrow planting system and this was assumed to be related to the deeper planting of corms in the former (34).

3-2-2 Moisture

Water requirements of saffron have been dealt with in a separate chapter in this book (chapter 5). However, a few points are considered here. Saffron is well adapted to the rainfall pattern of Southern Khorasan. In other words, its growth starts with beginning of rain in autumn and the vegetative growth ends by the termination of rainfall in spring. Water provided by rainfall is effectively used. On the other hand, since water requirements for other plants are low in winter, there is no competition with saffron for irrigated water at this time. However, initial irrigation in the fall and early summer, which are the most crucial ones for saffron flowers to emerge, are based on competition with other crops. In such cases priority is given to saffron due to its high economic return. Although water requirement is low in saffron, water stress affects the yield, growth and development. As is shown in Table 3-1 it appears that water deficit reduces economic and biological yield (34).

Table 3-1   *Effects of methods of irrigation and amount of water on economic and biological yield of saffron (34).*

Methods of irrigation	Amount of irrigation	Economic yield		Dry matter yield (kg/m²)
		*Dastah	*Sargol	
Flood irrigation	100% actual evapotranspiration	1.33 a	0.42 a	0.24 a
	75% actual evapotranspiration	0.99 b	0.69 b	0.16 c
	50% actual evapotranspiration	0.87 c	0.59 c	0.11 cd
	Rainfed	0.34 g	0.23 f	0.09 d
Furrow irrigation	100% actual evapotranspiration	0.79 d	056 c	0.29 a
	75% actual evapotranspiration	0.69 e	0.49 d	0.29 a
	50% actual evapotranspiration	0.49 f	0.35 e	0.18 b
	Rainfed	0.13 h	0.09 g	0.11 cd

*These are Persian terms for the whole stigma and pistil (Dastah) and the stigma alone (Sargol) which are based on quality attributes.

3-2-3 Soil

Saffron can be grown in a wide range of soils, with moderate structure and well infiltration, clay, siliceous, ferrugenous and gypsyferous soil are preferred. Since with soils rich in calcium, high amount of organic matter is facilitated, this type of soil is recommended (36). Barshad (6) claims that deficiency of calcium carbonate in soils may have a negative impact on productivity of saffron. Calcium carbonate has been reported (16) to facilitate availability of trace elements for saffron.

Soils with high moisture content and flooding condition are not suitable for saffron production, due to corm decomposition (21). Nutrient uptake in saffron is high (2). Each kg of total dry matter of saffron removes 12 grams N 3 g P and 22 g K from the soil (24). Shahandeh (33) found that 16 to 80% of yield variation in saffron could be attributed to edaphic factors and the determinant soil factors in order of importance were, soil organic matter, available P, N and exchangable K. Source of nutrients has also been shown to be important. Ammonium N has a negative while nitrate N has a positive impact on flower yield of saffron.

Saffron growers all over the world believe that saffron could not be grown in the same soil continuously. The scientific basis of this claim is not clear. However, some works have been conducted on this topic (15, 28). Gharaii and Bygi (15) studied different criteria of soil under saffron in the Estahban-Fars province of Iran for a long period and found no differences between soils under cultivation. Gharaii and Rezaii (16) found no differences on the amount of micronutrients in soils under saffron compared with other type of soils (Table 3-2).

Mycorrhizal studies in Gonabad and Ghaien showed the presence of VAM fungi on roots of saffron (24). Number of spores per 10 g soil was between 13 and 52 with the highest occurrence in December-January. In this study inoculation of corms with spores enhanced growth of corms.

Table 3-2 *Micronutrients in soils with and without saffron cultivation (16).*
            (mg/kg soil)

	Mn	Fe	B	Co	Cu	Zn	Cl
Year 1 of cultivation	2.15	0.80	196.0	0.01	0.36	0.22	142.0
Year 5 of cultivation	4.68	1.38	138.8	0.01	0.36	0.32	184.5
Year 10 of cultivation	2.67	0.92	312.4	0.1	0.36	0.16	138.0
Five years after one cycle of plantation	3.14	1.55	370.3	0.14	0.76	0.30	195.0
Twenty year after one cycle of plantation	7.40	2.91	486.0	0.01	0.90	0.20	114.0
Soil without saffron	4.61	1.47	486.0	0.14	0.63	0.33	315.5

3-3   PHENOLOGICAL STAGES

Phenological stages have been quantitativelly evaluated for different cultivated plants (19). However, this has not yet been developed for saffron due to its rather strange type of growth, e.g. some parts of development of saffron take place under the soil surface. Suggestions on these aspects have been made by different authors (8, 32, 5).

3-3-1   Phenological Stages on the Basis of Above Ground Parts

In general, growth stages of saffron could be classified in the following phases:

A—Generative phase

This phase starts with onset of cold weather in fall and is an important stage for growers. The main stimulating factor in this phase is irrigation in late summer and early fall. Although apparently, this phase starts with application of the first irrigation and emergence of the first flowers and ends with termination of flower emergence (15-25 days), its physiological process start well before the apparent flower emergence (6, 32).

Heterogeneity and lapse in flowering time could be associated with difference in homogeneity in corm development, depth of planting and soil physical criteria such as surface roughness. Soil crust has a negative impact on flowering characteristics of this plant (8).

B—Vegetative phase

This phase is the longest period in the life cycle of saffron and starts immediately after the flowering stage. However, when the initial irrigation starts earlier than the proper time, leaves may emerge simultaneously with flowers. This is not a proper practice because leaves could interfere with the picking of flowers. This phase lasts for at least 6 months (November to April). At this stage leaves are developed and provide necessary nutrients for corms. Cultural practices such as control of pests and diseases take place at this stage (8, 27). In Fig 3-1 the pattern of flower development is shown (26).

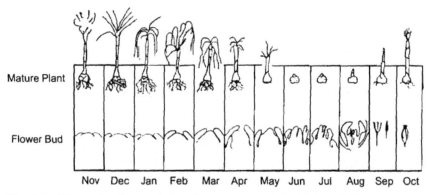

Fig. 3-1   Flower development of saffron over a period of 1 year (26)

C—Dormant phase

This phase starts with leaf withering and senescence in the spring and ends with the first irrigation in late summer and early fall. This period lasts for five months. Since there are no cultural practices in this period it has minor significance for the growers. This period is considered by the growers as a rest period of corms for flower production in fall (27, 30). Further classification of these phases based on leaf dimensions, number of leaves and leaf area could be considered as a tool for a quantitative analysis of phenological stages of saffron.

3-3-2 Phenological Stages based on Corm Development

Corm development takes place under the soil. Economic organs initiate and develop under the soil surface (28, 29). Therefore, this phase is associated mainly with corm development. Longevity of corms is normally low and after flowering and development of new corms (daughter corms), main corms deteriorate. Therefore, development of new corms starts with initiation of new buds and ends by production of mature corms (5, 23, 27).

On the surface of each mother corm there are several meristemic points (eyes), which are the base of buds for new corms. The number of eyes in a corm has been recorded to be up to 10 (28).

Activity of these eyes starts after termination of flower emergence in mid-autumn. These eyes are located at the uper part of mother corms (5, 12, 31). Activity of these meristemic points take place in two phases of slow and fast cell division based on the number of cells being divided in proportion to the total cell (Mitotic index) (8, 27). The slow phase starts after flowering and continues till early March. After this stage, fast division starts and continues up to withering and senescence of leaves in April and May. At this stage daughter corms are formed and developed. The number of corms produced in a mother corm negatively correlates with its size. Small corms have difficulty in producing flowers (12, 28). In contrast to the general belief that during summer, corms are in rest, new corms are active and flower development takes place at this time. Koul and Farooq (25) produced microscopic cuts of corms 13 times at 15-day and 4-day intervals in spring and summer and examined terminal meristems for 2 years. Based on their studies, the period between corm development and enlargement, from May to July, no meiotic activity in the cell was observed and, therefore, this period was considered as a dormant phase. Thereafter, initiation and differentiation of leaf cells begin by appearance of double ridges on both sides of buds. This process of change and development continues till early August and meanwhile leaves are formed. After this stage, by expansion and widening of upper parts of meristem, initiation of flowers start and sepals, petals, anthers and ovary are formed.

Initiation of flowers takes place from early August for around 20 days. It has been observed that, if corms are exposed to low temperature and moisture before this date, flowers do not develop and this is a good indication to conclude that flower initiation takes place only in this particular period (30).

Three types of roots are distinguished in saffron:

1. Absorbing roots which appear at the base of mother corms. These roots are fine and long (28).

2. Short roots which are thick and short and located individually in the base of forming daughter corms. These type of roots are formed in corms adjacent to the soil surface and, therefore, play an important role in connecting the daughter corms to deeper depth of soil for water and nutrient uptake. The stimulating factor for formation of these roots is probably light and temperature variation (28).

3. Connecting and uptaking roots which are formed in mother corms, near the differentiating daughter corms. These roots are thinner and longer than short roots and are formed roughly three weeks after formation of short roots. Therefore, their formation is not only dependent on light and temperature variation but presence of short roots are also important for their formation (28).

3-4 PHYSIOLOGICAL CHANGES DURING CORM DEVELOPMENT

In saffron, on an annual basis, seven months from late October to late April is the period for leaf growth and corm development, two months from late April to late June and early July is a dormancy period and three months is for flower initiation and development (1, 14, 28, 30). Corms act as a source of nutrients for flowers and newly formed corms. Physiological changes during the dormancy period of corms were studied by Milyaeva and Azizbekova (27). They stated that after flowering time, meristems of daughter corms, in upper parts of mother corms, become active. Afterwards, during autumn and winter, they propagate slowly and in late winter they grow fast. After vegetative growth, when leaves wither and dry out, mother corms deteriorated completely and nothing is left except the outer decomposed layer. In this stage corms transfer from a vegetative phase to a generative phase (Fig. 3-2). During the summer period flowers are initiated and developed.

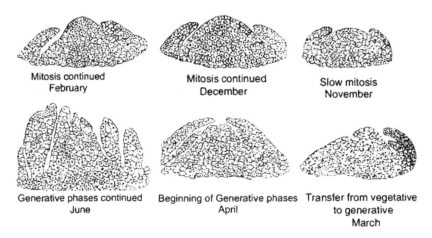

Mitosis continued
February

Mitosis continued
December

Slow mitosis
November

Generative phases continued
June

Beginning of Generative phases
April

Transfer from vegetative
to generative
March

Fig 3-2 Trend of corm development in dormancy phase (26)

Azizbekova and Milyaeva (5) found that use of gibberllin on corms in June stimulated flower meristems. In another experiment (11), gibberlic acid (GA) and Naphtaline Acetic Acid (NAA) were used on corms (100-500 ppm GA and 100 ppm NAA) in June. It was found that by increasing GA concentration, number and weight of flowers increased and flower emergence accelerated, whereas NAA delayed the activation time and therefore, no flower was produced. They also found that carbohydrate reserves, particularly starch in corms accumulated till June and thereafter, decreased and sucrose, which is a transferable sugar, increased.

Farooq and Koul (13) studied inhibitory compounds in corm under the dormancy phase and found that inhibitory compounds, especially ABA plays a major role in corm dormancy. It was also noted that GA_3 at concentration as low as 50 ppm stimulated corm activity. In another experiment these authors (25) applied Gibberlline and Kinetine to the base of corm twice in February and June. They noted that application of these compounds caused vigorous growth with longer roots, leaves and more flowers. Greenbery (18) treated corms with Gibberlline and found that although this treatment caused upper corms to be larger but the number of corms, compared with the control was lower. This was speculated to be due to more apical dominance caused by Gibberlline. Irradiation by gamma rays on corms has been found to increase number of corms, number of flowers and stigma weight in three consecutive years (3).

One of the main problems in saffron is heterogeneity in anthesis emergence which normally takes place during 2-3 weeks. This process caused uncertainty in the number of labor required for harvesting on a daily basis. An experiment was conducted by applying GA with concentration of 25 ppm to large corms of similar age for 48 hours and exposure to temperature ranging from 0-2°C for 72 hours. Results showed no changes in physiological and cytological processes of saffron (9). This may be due to the fact that flower initiation had taken place before application of these treatments.

3-5 GROWTH PARAMETERS

Apparent growth period for saffron is approximately 7 months and at this time leaves act as assimilatory resources for corms and roots. The amount of assimilates transferred to these organs depends on photosynthetic leaf area and efficiency of photosynthesis per unit of leaf area. Photosynthetic rate in saffron has not been investigated but photosynthetic area has been studied (34).

Leaves in saffron are in a vertical state (orthotopic) during emergence in the fall and change to a horizontal position (plagiotopic) as they develop and finally wither and senesce.

Photosynthesis in this type of spatial arrangement of leaves is not high. This could be associated with the fact that, at the early stages of growth, when there is not much competition between leaves for light, a horizontal orthotropic arrangement of leaves is advantageous. However, with development of leaves an erectus arrangement is preferred. This trend normally happens in many plants such as wheat and rice. Therefore, a reverse temporal trend of leaf arrangement for saffron causes light not to be intercepted effectively and, hence, a low photosynthetic efficiency. Another criterion of saffron leaves is the white color of the main vein which comprises a good proportion of the leaf area. Lack of pigments in the leaves could be another cause of low photosynthetic efficiency.

If there was variation between population of saffron in terms of leaf length, genotypes with smaller leaves and higher number of leaves are preferred, because these types of leaves remain in vertical state with higher photosynthetic activities. Some environmental factors cause a reduction in leaf length, but this reduction is not normally compensated by production of more leaves, and hence a reduction in leaf area index is caused. In Table 3-3 leaf length and dry matter in response to fertilizer is shown for an eleven-year old saffron field in Mashhad (30). As is seen, lack of some nutrients causes reduction in leaf length and dry matter.

Table 3-3 *Effect of fertilizer on leaf length (cm) and dry matter (ton/ha) in an eleven-year saffron field in Mashhad (29)*

Treatment	Manure	NPK	NP	N	NK	P	PK	K
Leaf length	30.6 ab	34.0 a	33.0 ab	27.0 b	26.8 bc	17.0 d	17.2 d	20.6 d
Leaf dry matter	5.1 ab	7.19 a	6.8 a	3.8 abc	2.8 bc	1.6 c	2.0 c	1.5 c

N was provided by Urea 46%
K was provided by Potassium Sulphate 50%
P was provided by Superphosphate 48%
All at a range of 50 kg/ha individually or in combination
Manure was semi-decomposed cow manure (25 ton/ha)

Optimum Leaf Area Index (LAI) for saffron has been reported to be low and does not exceed 1.5 (34). This figure varies from 4-5 for wheat, barley, rice and cotton. Shirmohamadi (34) evaluated LAI for saffron in response to irrigation treatments (Table 3-4). He found values of 0.23 for LAI, 31 days after the initial irrigation to 1.24 when measured 173 day after the initial irrigation.

Table 3-4   *LAI for saffron in response to irrigation treatment (34).*

Days after initial irrigation	Flood irrigation			Rainfed
	100% ET	75% ET	50% ET	
31	0.23	0.21	0.13	0
60	0.51	0.50	0.39	0.14
89	0.62	0.53	0.41	0.27
110	0.67	0.56	0.53	0.28
173	1.24	0.96	0.84	0.52
202	0.65	0.62	0.59	0.18

He proposed calculation, rather than measurement of LAI based on a strong correlation which existed between dry matter and leaf area. Figure 3.3 shows this correlation with $R^2 = 0.884$ and A = 31.797, W + 112056

where   A = leaf area (cm^2) and W = leaf weight (g)

        ET = evapotranspiration.

3-6 BIOLOGICAL YIELD

Biological yield in saffron includes mainly leaves and corms. The amount of leaves produced per unit of land depends on the age of the field and availability of mineral nutrients. The older the age of the field, the more leaves are produced per unit of land because of relative crowding of the crop. Yield of corms in response to fertilizers is shown in Table 3-5 (29) for an eleven-year old field.

In another study (34) it was found that the minimum dry matter yield of leaves was 1355 kg/ha with furrow irrigation and 1301 kg/ha

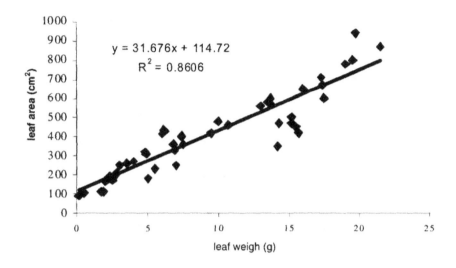

Fig. 3-3 Correlation between leaf area and leaf dry matter (34)

Table 3-5 *Effects of fertilizer on weight (ton/ha) and number of corms (× 1000) (29)*

Treatment	Manure	NPK	NP	N	NK	P	PK	K
Corm+ tonica	35.8	31.9	3.9	25.6	24.3	15.8	12.8	14.2
Corm	22.2	21.3	19.2	14.5	13.9	8.7	8.2	8.1
tonica	13.6	12.5	12.7	11.1	10.4	7.1	4.6	6.1
No. of corms/ha	7100	6368	5583	4975	5538	3510	2623	2225

with flood irrigation. Yield of corm were 15.3 ton/ha and 20.3 ton/ha respectively, and the number of corms for these treatments were 3192×10^3 and 2228×10^3 per ha respectively.

It should be noted that dry matter proportion of corm is 40% and that of leaves is 93% On the other hand all leaf biomass produced is for the same year, whereas in the case of corm, a good proportion is from the previous year (30). Furthermore about 40% of corm weight is composed of covering dead material (29). Above all, not all the corms produced are suitable for plantation because small corms do not produce flowers in the first year (23, 30). Corms with more than 5 g weight are capable of producing flowers in the first year (15).

Table 3-6 *Dry matter yield of leaves, weight of corm and number of corms per hectare in a four-year old field in Shiraz (34)*

Irrigation methods	Percent of water provided	Leaf dry matter kg/ha	Corm weight (ton/ha)	Corm number (000's)
	100%	1301	20.3	2228
Flood	75%	812.5	16.7	1993
irrigation	50%	553.1	10.5	1366
	Rainfed	173.1	4.95	625.9
	100%	1355	15.3	3192
Furrow	75%	1276	11.2	2284
irrigation	50%	770.6	10.43	1514
	Rainfed	289.1	4.97	1255

In another experiment (30) it was shown that corms less than 8 g were not recommended. Proportion of corms in a population based on the size has been investigated (Table 3-7) in Shiraz (34).

Table 3-7 *Weight and number of corms in a four-year old field (34)*

Corm size	Weight kg/ha	Proportion (%)	Number 1000/ha	Proportion (%)
>8 g	12700	62.19	965	42.61
4 to 8 g	4520	22.13	595	26.27
<8 g	3200	15.67	705	31.12

Small corms have other uses such as Dextrin production due to their high starch content.

3-7 ECONOMIC YIELD

Economic yield is normally measured in terms of flower yield or dried saffron stigma per a unit of land (ha). Since dried saffron is composed of stigma plus style (which may be called stigyle), in some cases it is measured on the basis of each component separately as sargol (stigma) or dastah (stigyle). Flower yield is much lower compared with leaf or corm yield. Yield of saffron (stigyle) is the product of the following components:

$$Y = A \times B \times C \times D \times E$$

where,

-   $-Y$ = dried saffron spice (g/ha)
-   A = number of flower producing corms
-   B = number of flowers per corm
-   C = weight of a flower
-   D = proportion (%) of style + stigma (stigyle)
-   E = percent dry weight of saffron spice.

Suppose, the number of flower producing corms are 1,000,000 per ha, the number of flowers per corm is 2, weight of a flower is 0.35 g and proportion of style plus stigma is 8% and dry matter content of saffron (stigyle) is 15-19%, yield per ha will be 10.40 g based on this formula. However, variation of yield in saffron is high and depends on different factors such as the age of the saffron field, management practices and environmental conditions. Mean yield of up to 13.5 kg/ha for an eleven-year old field has been reached (29). Yield of saffron depends on the following factors:

-   Age of the field
-   Size and number of the corms planted
-   Timely irrigation
-   Timely harvest
-   Control of pests, diseases and weeds.

Yield variation in saffron associated with age of the field has been shown in Table 3-8.

Table 3-8 *Yield of saffron based on the age of saffron field*

Age	Yield of flowers (kg/ha)	Yield of dried saffron (kg/ha)	Reference
3	600.2	6.910	9
1	-	4.00	27
2	-	13.30	27
3	-	20.00	27
4	-	13.3	27
Mean of 11 years	850.0	13.48	10
1	4000.0	13.0	36

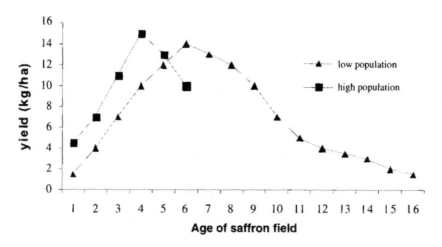

Fig. 3-4   Yield variation on the basis of age and plant density (24, 27)

Table 3-9   *Yield components for saffron*

Number of flowers (per kg)	Weight of flower (g)	Weight of fresh saffron (g/kg flower)	Weight of dried saffron (g/kg flower)	Weight of stigma (g/kg flower)	Weight of style (g/kg flower)	Reference
2500	0.3-0.5	87.5	17.5	-	-	37
2500	0.3-0.5	80.0	16.0	12.0	4.0	28
2173	0.46	76.32	12.74	9.48	3.26	14
-	0.4	82.78	15.86	-	-	10
-	-	-	11.50	8.05	3.45	9
2350	-	-	14.46	9.04	5.42	2
-	-	100.0	14.28	-	-	4

In Table 3-9 yield components of saffron is shown.

3-8   SUMMARY

Saffron is a plant with a special ecophysiological character, somehow different from other plants. Plant development and economic yield of saffron is related to photosynthetic reserves in the corms from the previous growing season. Leaves are active from November to April and photosynthetic materials are stored in the corms for production of new corms and flower initiation. LAI of saffron is as low as 1.24, leaves do not stay in a vertical position during some part of the growing season, wither, and therefore spread over the ground. This together with the white color of the main vine of the leaves cause low photosynthesis efficiency. Since most of the physiological process of saffron takes place under the ground, temperature and moisture conditions of corms in soil play an important role in plant development.

References

1. Abrishami, M. H. 1997. Iranian saffron: historic, cultural and agronomic prospects: Astan Ghods Razavi Publishing Co. (Persian).

2. Ait-aubahou, A., and M. El-otmani, 1999. Saffron cultivation in Morocco, in: "Saffron" (M.Negbi, ed.), Harwood Acadamic Pub. Amsterdam. Pp. 154

3. Akhund-Zade, I. M. and R. Sh. Moaferova, 1975. Study of effectiveness of gamma irradiation of the saffron. Radiobiologica, 15: 319-322.

4. Amirghasemi, T. 2001. Saffron, red gold of Iran. Ayandegan Publishing Co. (Persian).

5. Azizbekova, N. Sh., E. L. Milyaeva, N. V. Lobova and M. Kh. Chailakhyan. 1978. Effect of gibberllin and kinetin on formation of flower organs in saffron. Soviet Plant Physiology. 25 (3): 471-476.

6. Barshad, I, E. Halevy, H. A. Gold and G. Hagin. 1956. Clay minerals in some limestone soils from Israel. Soil Sci. 81: 423-437.

7. Behnia, M. R. 1991. Saffron agronomy. Tehran University Press.

8. Biswas, N. R., S. P. Patta, Raychaudhuri, and C. Dakshinamurti. 1975. Soil condition for the growth of saffron of pampoye (Kashmir). Indian J. Agric. Sci. 27 (4): 413-418.

9. Chrungoo, N. K. and S. Farooq. 1984. Influence of GA and NAA on yield and growth of saffron. Indian J. of Plant Physiology. 27: 201-205.

10. De Mostro, G. and C. Ruta. 1993. Relation between corm size and saffron flowering. Acta Horticulture. 344: 512-517.

11. Farooq, S, and K. K. Koul. 1983. Changes in gibberllin-like activity in corms of saffron plant (*Crocus sativus L.*) during dormancy and sprouting. J. Plant. Biochem. 178: 685-61.

12. Berengi, A. G. 1992. Physiology of growth and development of saffron. Ziton Scientific Magazine. 108:36-38(Persian).

13. Filabi, A. 1999. Yield component of saffron. Proceedings ot the Second National Symposium on Saffron and Medicinal plants. 8-9 November 1994, Gonabad, Iran.

14. Gharaii, H. A. and M. Beygi. 1991. Evaluation of physicochemical and mineralogy of soils in saffron growing areas of Estahban. Scientific and Research Organization of Iran Shiraz Institute, Annual Report.

15. Gharaii, H. A. and A. R. Rezaii. 1993. Effect of saffron cultivation on trace elements of soil in Estahban. Scientific and Research Organization of Iran, Shiraz Institute, annual report.

16. Goliaris, A. H. 1999. Saffron cultivation in Greece, in "Saffron" (M.Negbi, ed.) Harwood Academic Pub. Amesterdam. Pp. 154

17. Greenbery-Kaslasi, D. 1991. Vegetative and reproductive development in saffron (*Crocus sativus L.*) M. Sc. Thesis. Ten Heberw University, Jerusalem.

18. Haun, J. R. 1973. Quantitative wheat growth stages. Agron. J. 65:116-119.

19. Housaini, M. 1997. Effect of foliar application of nutrients on yield of saffron. Scientific and Industrial Research Organization of Iran, Khorasan Institute, Annual Report.

20. Ingram. J. S. 1984. Saffron (*Crocus sativus L.*) Tropical Science. 11:1771-1774.

21. Javanmard, S., G. Ahmadian, Sh. Malbosi and D. Dashtiani. 2002. Evaluation of risk for saffron production in South Khorasan. First Saffron Festival. 2-3 December 2002, Ghaen, Iran,

22. Kaushal, S. K. and R. G. Mpadhay, 2002. Studies on variation of corm size and its effect on corm production and flowering in *Crocus sativus L.* in Himachal Pradesh. Research on Crops. 3(1):126-128.

23. Kianmehr, H. 1994. Endotrophic micorehyza of saffron in Khorasan and its possible application. Proceedings of the Second National Symposium on Saffron and Medicinal plants. 8-9 November 1994, Gonabad, Iran.

24. Koul, K. K. and S. Farooq. 1984. Growth and differentiation in the shoot apical meristem of saffron plant. Journal of the Indian Botanical Society. 63:153-169.

25. Madan, C. L., B. M. Kapar and U. S. Gupta. 1966. Soffron. Econ. Bot. (20)4.

26. Milyaeva, E. L. and N. Sh. Azizbekova. 1978. Cytophysiological changes in the course of development of stem apex of Saffron. Soviet. Plant Physiol. 25:227-233.

27. Negbi, M. 1999. Saffron (*Crocus sativus L.*) Harwood Academic Publishers.

28. Sadeghi, B. 1989. Effects of chemical fertilizer, and animal manure on corm, leaf and saffron yield. Scientific and Research Organization of Iran. Khorasan Institute, annual report.

29. Sadeghi, B. 1983. Effects of corm weight on flower initiation of saffron. Scientific and Research Organization of Iran. Khorasan Institute, Annual Report.

30. Sadeghi, B. 1996. Effects of corm storage and time of planting on flower initiation. Scientific and Research Organization of Iran. Khorasan Institute, Annual Report.

31. Sadeghi, B., S. A. Aghamiri and A. K. Negari. 1997. Effects of irrigation on yield of saffron. Scientific and Research Organization of Iran. Khorasan Institute, Annual Report.

32. Shahandeh, H. 1990. Evaluation of physico-chemical characters of soil and water for saffron production in South Khorasan. Scientific and Industerial Research Organization of Iran Khorasan Institute, Annual Report.

33. Shirmohamadi, A. and Z. AliakbarKhani. 2002. Effects of methods and number of irrigations on Leaf Area Index, canopy temperature and yield of saffron. MSc. Thesis, Shiraz University.

34. Tammaro, F. 1999. Saffron in Italy. In "Saffron" (M. Negbi, ed.) Harwood Academic pub. Amesterdam. 154pp.

35. Vafabakhsh, G., H. Ahmadian, D. Shibani, and G. Badagh Jamali. 2002. Potential climatic zone for saffron growing in Iran. Poceedings of the second. National symposium on Saffron and Medicinal plants. 8-9 November 1994, Gonabad, Iran.

36. Valizadeh, R. 1989. Evaluation of saffron leaf as animal feed. Scientific and Research Organization of Iran. Khorasan Institute , annual report.

CHAPTER 4

Saffron Production Technology

A. Molafilabi

Iranian Scientific and Industrial Research Organization, Khorasan Center

4-1 INTRODUCTION

Saffron has been regarded as a native plant of the Iranian plateau (2) and has been cultivated in the country for centuries. Saffron is an important cash crop in Khorasan and has been a major revenue generating product for small farmers in water deficit areas of the south Khorasan (6). This crop has a specific criterion such as low-water demanding, water requirement at periods of high water availability, ease of transportation, low demands for expensive machinery and labor requirements at the time of harvesting and therefore job generation for local people, all of which are suitable for areas with low industrial development potential (6). Although this crop has a long history of production in the southern Khorasan, not much progress is made in developing production and processing technologies. Even in countries such as Spain, India and Italy not much advancement has been made with saffron production (33).

4-2   CROP ROTATION

Rotation is crucial for control of pest and diseases and also enhancement of soil fertility (21). There is evidence to show (27) that sugar beet, potatoes and alfalfa are not suitable crops for rotation with saffron. Farmers have experienced that saffron should not be cultivated on the same land and a proper fallow period should be practiced or bring other crops such as cereals or pulses in rotation (10). In Spain a rest period of 10 to 20 years is practiced for soils presently under cultivation of saffron (15).

4-2-1   Multiple Cropping

In some part of Khorasan, saffron is cultivated as alley cropping between rows of orchard plants such as grapes' barberry and almond. This is also practiced in Kashmir, where saffron is planted between 6 × 6 m rows of almond. Intercropping saffron with black cumin has been introduced recently. These two plants have similar criteria such as perenniality, dormancy habit and underground corm. In Spain saffron has also been planted between rows of olive plants and in vineyards (32). Planting saffron with ornamental plants such as rose has also shown promising results.

4-3   CULTIVATION OF SAFFRON

4-3-1   Soil Preparation

Medium textured soils with a good natural drainage potential, fairly deep and smooth surface area with no salinity is preferred for saffron planting. Soil is ploughed in autumn or winter and animal manure of 20 to 100 tones per hectare is applied. In Spain a deep ploughing of 25-35 cm is practiced in winter and the fields are harrowed to eliminate weeds. A second shallow tillage is carried out in early April and this may be repeated in a later stage and finally animal manure is applied. In Kashmir soil preparation is practiced in February and

March with 3 to 4 tillage operations and application of 5 to 6 tons per hectare of animal manure. A last tillage is practiced in late April and the final seed bed leveling is carried out.

4-3-2   Planting Methods

Saffron is planted either in dry or moist beds (27). In traditional systems, corms are planted in hills 25 cm apart with sometimes up to 15 corms per hill with no row arrangement. When corms are planted in rows, shallow ditches 30-35 cm apart are made by a furrower and corms are arranged in hills of 3 to 15 corms and finally covered with soil (10). Flat bed planting has been reported (8) to be advantageous compared with furrow planting. However, due to shortage of man power, mechanized planting may be required in the future (29, 25). Raghimi (29) made some modifications on a potato planter to make it suitable for planting of corms and found that the time required to plant one hectare was 4 hours. He reported some of the advantages of mechanized planting as follows.

1. Year 1. Since in mechanized planting 4-5 corms are located in each hill, total corms needed for each hectare is 3 tons. Uniformity in planting depth (25 cm) is also achieved in this method and corms are covered with soil by rare disks on the implement.

2. Soil crusting is avoided in row planting and there is no need to break the crust as it is practiced in flat plantings.

3. Saving in labor costs. Labor force required for planting one hectare is 80 person-day, while time required for this operation in a mechanized system is only 4 hours.

4. Picking flowers and also other management practices are easier in mechanized planting compared with the traditional system.

5. Since plants are not in direct contact with water in furrow plantings, spread of disease is also reduced.

In an attempt to design a corm planter for saffron, Saiidirad et al. (39) proposed a prototype with ability to plant corms in rows of 20 cm apart and 7-15 cm distance within rows with a planting depth of 15 cm. In their experiment they evaluated some physical criteria of corms such as form, size, actual and apparent specific weight and also angle of friction of corms with the container of the implement. Based on these criteria a single-corm planter with all required devices and transmission power was designed (Fig. 4-1).

Fig. 4-1   A view of corm planter

In all mechanized designs the important point is the proper placement of corms in soil in such a way that the base of the corms is in direct contact with the soil and the growing points are in an upward position. Corms are also planted by furrow planting operated by small orchard tractors (10). In such cases after furrows are made corms are located in the furrows by labor in such a way that the bases of the corms are in contact with the bottom of the furrow and the growing points are located upward. Finally alternate furrows are made by which the soil is directed on the corm-containing ditches and therefore proper coverage of corms is achieved. The new furrows are used for irrigation practice. Planting depth of 15-20 cm with rows up to 35 cm apart can be used for mechanized furrows. Planting corms in moist beds requires irrigation of the field a few days prior to planting. In such cases after preparation of the soil corms are planted in hills on the rows 25 cm apart and 15 to 20 cm distance between each hill (2, 6). In Spain after preparation of soil ditches of 20 cm deep are made and corms are located in the bottom of ditches in two rows with 8 to 10 or 12-15 cm apart depending on the arrangement of location of the corms, either rectangular distance or alternate (zigzag) (Fig. 4.2). Corms are then covered with the soil from the adjacent furrow which is 30-35 cm apart (32, 33, 35).

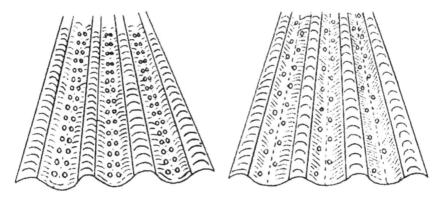

Fig. 4-2 Two different types of arrangement of corms in furrows

In France saffron field are rejuvenated every three years (14). In Greece row planting is practiced with 20-25 cm apart and 11-13 cm distance between corms in the rows and a planting depth of 15-17 cm (14, 20). In Italy annual planting is operated and each year corms are uplifted and replanted in depth of 8-10 cm with 1-2 cm distance between corms in the rows (41). In Morocco saffron is planted in hilly areas on traces with hill planting of 2-3 corms per hill on rows of 20 cm apart with 10-15 cm between each hill and at depths of 15cm.

4-3-3 Selection of Corms

Saffron is solely reproduced by corms (2, 10, 15), therefore selection of proper corms is crucial for production. Large corms with no injury from 2 to 4 year old fields are preferred. Size of corms is very variable and ranges between 1 and 20 gr (10). Sadeghi (36) in evaluating the effect of size of corms on flowering potential found that corms size up to 2 g weight have no flowering potential and up to 8 g their potential is limited. However corms with 10 g or more are productive and those with 14 g weight produce flower in the first year with a yield of up to 3.5 kg saffron per ha. Large corms produced more and larger corms in two following consecutive years too and therefore, productions of higher yield of 11 and 20 kg/ha for the second and the third year respectively. In this experiment the average yield for 3 years was twice as much as that for the conventional fields in the area.

In another trial (22) it was noted that the lowest corm size for production of flower was 6-5 g for the first year and 7.5 g for the second and the third year. There was no relationship between the corm size and the time of flower emergence. However, large corms produced more flowers. In the same trial it was noted that corms with 22.5 g weight produced 2-3 flowers per corm in the first year those with 23.5 g produced 3-4 flowers in the second and those with 26.5 g produced 2-3 flowers per corm in the third year. It is therefore apparent that only the size of corms up to a particular point increases the flowering potential and not necessarily thereafter. This could be associated with the initiation of aging process of the corms (22, 27).

It must be noted that production of such sizes for corm is very rare and based on our experiences it is not usual to get corms more than 15 gr in the field under dry bed corm lifting. In such cases corms can be classified in 4 categories of 2-4 g (small), 4-6 g (medium), 6-8 g (large), more than 8 g (very large). However under moist bed corm lifting, where irrigation is applied before lifting the corm, corm size is increased up to two folds. This is not a normal practice and in such cases corms are transferred to the new beds as soon as they are lifted.

Corms with 2.5 cm diameter (6 g weight) have been recommended for planting (24). In Spain large corms from 4-year old fields are selected for planting (32).

It seems, age of the field is an important factor in size of the corms. Therefore, it has been recommended (36) to reduce the age of the field which is normally 8-10 years in traditional system in Iran to 4 years in order to avoid reduction of size of corms which have a negative impact on flowering potential.

Age of saffron fields in India is 6-8 and in Spain 4-5 years (15, 32, 35, 19). Highest yield of saffron in Iran is normally obtained from the third year, therefore it appears that with increasing age, the crowding of corms causes reduction of their size and production of more corms with smaller size. It is therefore, recommended to lift the corms for replanting not later than four years. This practice is normally done on an annual basis in Italy (41).

4-3-4   Time of Corm Lifting

Corms are lifted in dry or wet bed. However, dry bed is preferred because corms remain dormant during the summer months (10). In wet bed, corms are transferred directly to the new field for an immediate planting. Since corms are dormant from April to June, lifting is recommended in June (6, 10, 29, 27, 16) for cultivating in the new field as soon as conditions permit. However, corms may be stored in dry and cool stores (3–5 °C) for some time. Storing corms may reduce flowering potential and is not recommended (6, 10, 11, 18).

4-3-5   Time of Planting

Corms are planted from the time of leaf senescent in late May till early October (6, 10, 27, 35). In Iran this is normally carried out in August to early October, but in some case early planting in June and July may be carried out. Planting in the hot months of summer may cause desiccation of corms and therefore is not recommended (6, 10). In Spain planting is done in April and June (15, 20, 32) and in Kashmir in July and August (15).

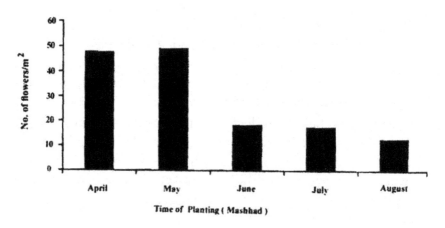

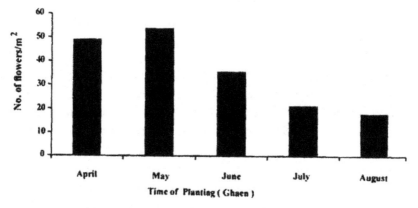

Fig 4-3 Effect of time of planting on flower production in the first year in Mashhad and Ghaen.

Sadeghi (35) in his trial on the effects of storage and planting on flower production found that April till June are appropriate period for planting (Fig. 4-3). This has also been confirmed by others (12, 16, 27, 35).

4-3-6   Preparation of Corm for Planting

Large and healthy corms from young fields are selected and outer corm cover is removed and finally they are planted. It is recommended to apply fungicide on the corm before planting (10,

15, 27) and 5% sulfide of copper Benlit 1.5/1000 and acaricide Emayt have been recommended (32). Fungicides, Serezan, Teritizan and Geraminon, with concentration of 300-500 gr for 100 kg corms have also been recommended (6, 10).

4-3-7   Planting Density

Number of corms required per unit of land depends on the planting method and size of corms and also the habit of farmers and varies between 1.5 to 10 tons per hectare (6, 10). In Kashmir this amount is 8 tons per hectare (15) and in Morocco 3 tons per hectare (3). In Greece 230 to 250 thousand corms per hectare and in Spain 300 thousand corms are used (41). Alavi et al. (5) found that increasing plant density increased the yield and 50 plants per m^2 (10×20 cm) were recommended. Based on an optimum size of corm for planting, which is 8 gr, this plant density requires 4 tons corms per hectare (36). Ghalavand and Abdulahian (13) found 30×10 planting distance to be more productive than other pattern studied. In general findings in Iran indicate that 50 corms per m^2 with a size of 4–5 gr is an appropriate density (5, 8, 13 35, 36).

4-4   FIELD MANAGEMENT

4-4-1   Application of Fertilizer

In Iran cow manure with rates of 20–80 tons/ha based on the type of soil and farmers habit, is applied to saffron (1, 6). However, chemical fertilizers are used with rates of 100 kg/ammonium phosphate at the time of breaking the soil crust (after first irrigation in early autumn) and 100 kg/ha Urea at the time of first weeding (10). In Spain due to high soil organic matter only 10-15 tons/ha animal manure is applied (32). In India 15-22 tons animal manure per ha at the time of planting with 5-10-15 kg NPK is used. Shahandeh (40) found that soil organic matter, available phosphorus, mineral nitrogen and exchangeable potassium are important yield determinants. He also noted that NH4-N has a negative but No3-N has a positive effect on saffron yield.

Saffron is believed to be a low nutrient requiring plant and fertile soils with high nutrient contents causes excessive vegetative growth which is not advantageous for saffron. Animal manure with high potassium content is used for saffron. Torabi and Sadeghi (42) found that in Mashhad early February is a critical period of growth and development for saffron. They believe that although roots are active up to early February, they do not contribute to nutrient uptake because of deterioration of mother corms and, therefore from this point growth of daughter corms is dependent on the storage substances from the mother corms and the current photosynthitate.

Sadeghi, Razavi and Mahajeri (38) in Torbate Hydariah and Mashhad found that soil in Torbate hydariah with 1% organic matter require 50 kg N and in Mashhad where organic matter in soil is low (0.3%) application of 25 tons/ha of manure was the most effective treatment (Table 4-1).

Table 4-1 *Yield of flower and saffron (kg/ha) in response to fertilizer during 8 years in Torbat Haydariah and Mashhad (kg/ha)*

Location	Yield	N	P	K	Manure	NP	NK	PK	NPK
Torbat Haydariah	Flower	77.1	40.9	42.6	67.7	73.4	76.5	44.5	76.0
Mashhad Haydariah	Flower	72.8	29.1	31.3	88.5	76.5	75.9	29.8	79.9
Torbat Haydariah	Mean	10.6	5.16	5.37	9.32	10.1	10.5	5.92	10.4
Mashhad	Mean	10.0	3.7	4.0	11.8	10.5	10.4	3.8	11.0

Sadeghi (36) recommended application of 100 kg/ha Urea after picking the flowers. Higher rates of N fertilizer seem to stimulate more vegetative growth and hence lower flower yield. Behnia (9) found that under climatic conditions of Birjand, the highest yield of flower was obtained by application of 50 kg N, 50 kg P205 and 20 tons of animal manure per hectare. He also found that higher rates of N (100 kg N/ha) reduced yield of flower.

Rezaiian and Pasban (34) found that in Ferdows application of 100 kg Urea and in Gonabad 125 kg/ha showed the best result. Second application of 124 kg/ha sulphur coated urea after primary application of 100 kg N/ha also caused high yield. Housini (18) found that foliar application of aqueous fertilizer of 12% N, 8% P_2O_5 and 4% K_2O in combination with Fe, Zn, Mn and Cu chelates in early March with concentration of 0.007 increased yields by 33%.

In conclusion it is shown that saffron is a low nutrient demand plant and requires a modest amount of nutrients (18) and high application of fertilizers and in particular nitrogen fertilizer promotes vegetative growth and lowers the yield (18). Application of more than 100 kg urea/ha has shown to reduce yield (9, 23, 37, 40). Split application also showed unsatisfactory results compared with one application of complete fertilizer (18).

4-4-2 Breaking Soil Crust

Breaking soil crust is an important operation in saffron production. This is done after the first irrigation by harrowing or cultivator or other similar implements. Breaking crust in the soil surface at depth of 5-10 cm, facilitate flowers to emerge.

4-4-3 Weed Control

Weed management is an important practice in saffron production. Weeding is practiced after the harvest of flowers (second irrigation). This also helps impacted soil between rows to loose up. When required, second weeding is carried out one month later. For control of spring and summer weeds additional weeding may be needed. During the summer dormancy of corms light cultivators could be used (6, 24).

Reshed (31) found 184 species of weed in saffron fields in South Khorasan of which 20 species were dominant. These species were from 128 genus and 33 families of which 113 were annual and 71 perennial species. The most prevalent species were from Asteraceae (32 species), Fabaceae (24 species), Poaceae (21 species), Crucifereae (19 species), Chenopodiaceae (12 species), Caryophylaceae (10 species), Boraginaceae (10 species) and Ranunculaceae (17 species).

The most important genus were Descurainia (Gonabad), Hordeum (Ghaen), Bromus (Birjand), Convolvulus and Hordeum bulbosum. Rage, Mobayen and Faghih (28) evaluated weed flora of saffron fields in Estahban, and found that Hordeum was the most noxious weed in saffron fields and nearly 70% of weed control costs were for this weed. Broad leaf species including Descurainia, Muscari and Malva were also prevalent. They also showed that herbicides gallant with a rate of 2 liters/ha was effective in controlling Hordeum and showed no adverse effects on saffron. Mixture of basagran + fusilade and also basagran + gallant with rates of 2 and 4 liters/ha respectively were effective in controlling some weed species. Herati (17) found that preplant herbicide of sonalan controlled 15 species of 17 total species found in saffron fields (except Achillea and Descurainia). Amiri et al. (7) found that pre-emergence herbicide sencor was effective for controlling broad leaf weeds and gallant for controlling narrow leaf weeds after flower harvest. Rahimian (30) in his trial on herbicides application before emergence at the time of breaking soil crust in October, after emergence following flower harvest (November) and finally in March, found that application of sencor and sonalan as pre-emergence herbicides was the most effective weed control without any adverse effects on saffron plant. Abbasi (1) showed that sencor applied either as pre-emergence or post-emergence and also gallant and fusilade applied as post-emergence and sonalan as pre-emergence were the most effect herbicides for weed control in saffron. Princep and gesaprim have also been used for weed control in saffron at a rate of 10 kg/ha (14). However, hand weeding is the most effective way of weed control and it is safe and carried out by families and local labor, but it is costly. In general mechanical weeding and keeping saffron as an organic product is recommended. However, when it is necessary to apply herbicides, pre-emergence type at the time of breaking soil crust is recommended (1, 7, 17, 28, 30). In summer-time general herbicides such as round-up and 2, 4-D could be applied for control of grasses (10).

4-5 HARVESTING

Harvesting saffron includes picking the flowers and separating the stigma. Picking flowers starts as soon as they appear in the field. This is done on a daily basis because flowers are short-lived and if they are left for a longer period, not only can they be damaged, the quality of saffron also decreases (26, 29). Picking flowers begins from October to November in Khorasan and differs in the region according to the climate variability and time of first irrigation. Flowering period of a field lasts for 15 to 25 days, starting gradually and reaching a peak from the seventh to the tenth days. Flowers are picked early in the morning and before sunrise, while in some areas this is also done in the afternoon (2).

In Spain flowers are picked from mid-October to early November and in Kashmir it is done at the end of autumn. Picking flowers is carried out by hand and there is no mechanical device for this purpose (32, 26, 30). Separation of stigma, which is a delicate job, is done by hand soon after flowers are picked and carried home.

Yield of saffron depends on climatic and edaphic conditions and also management practices such as planting methods, weed control and size of corm at planting (15). Yield in the first year is low and increases in the following years (2, 6, 10). Average yield of saffron in

Table 4-2 *Comparison of saffron yield for Iran and Spain in different ages (kg/ha)*

Country	Year 1	Year 2	Year 3	Mean	Source
Spain (farmer fields)	4-6	10-12	16-18	10-12	47
Iran (farmer fields)	0-23	1.8	3.7	1.9	6
Iran (experiment)	2.5	11.7	20.3	11.5	20

Table 4-3   *Comparison of environmental and management practices for saffron in Iran and Spain (12)*

Condition or practices	Iran	Spain
Altitude	>1000 m	500-700m
Mean annual temperature	8-16 °C	13-17 °C
Relative humidity	40-50 %	60-85%
Annual precipitation	120-210 mm	412-453 mm
Soil texture	Variable	Medium to heavy
Soil organic content	<1%	>1%
First tillage for bed preparation	March-April	March-April
Sorting corms	Usually not done	Practiced
Corm size for planting	Small, medium, large	Large
Disinfection of corms	Usually not done	Practiced
Planting depth	15-20 cm	18-22 cm
Planting methods	Hills with 3-15 corms	Single corm planting
Distant between rows	25-30 cm	30-35 cm
Distant between plants	12-20 cm	5-8 cm in parallel rows
Time of planting	August – September	April-August
Irrigation	At least 5 times	Mainly rainfed and in irrigated fields only twice
Animal manure	Used differentially	10-15 tons/ha
Weed control	Mainly mechanical	Mainly mechanical
Diseases	Not a serious disease	Corm decay
Pests	Mouse and other rodents	Mouse and other rodents
Time of flower emergence	October to November	Late September to mid November
Harvesting methods	By hand	By hand
Flowering period	Nearly 20 days	Nearly 15 days
Flower transportation	In bags	In baskets
Separation of flowers	With hand (dastah) or Poshal	With hand sargol and dastah
Mean yield	4.7 kg/ha	8 kg/ha (mean rain fed and irrigated)
Age of fields	7 to 12 years	Usually 4 years
Drying methods	Free air and recently by heaters	By heaters
Time of drying	3-7 days	2-5 hours
Final moisture content	Nearly 7%	Nearly 13%
Packaging	To some extent	Practiced

Iran for a period of 25 years has been recorded at 4.7 kg/ha (4, 5, 9, 11). Maximum yield is obtained in the fourth and the fifth years (10, 27). Annual yield of 15 kg/ha and minimum of 3.8 kg/ha for some fields has been recorded (36). Yield can be improved by shortening the average age of saffron fields from 8 to 4 or 5 years (36).

Average yield of saffron in Spain has been recorded at 10-12.5 kg/ha, with 6-10 kg/ha for the first year and 16-18 kg/ha for the following year (15, 32). In Tables 4-2 and 4-3 comparative analysis on yield are shown for Iran and Spain.

Arial parts of saffron are a good source of animal feed and the leaves are used directly or after harvesting for animal feed (32, 43).

Corms are also a source of income for farmers and yield of corms in the fourth year reaches 4 times higher than the first year. Very small corms which are not suitable for planting can also be used as animal feed (4).

CONCLUSION

There has not been a pronounced change in technology of saffron production worldwide since many years and this plant is still grown manually and therefore is a labor-intensive crop. It has been recommended that for improving yield, selection of large corms is crucial (>8 gr). Plant density of 50 plants/m^2 with single corm and row planting has been a promising practice. Age of fields, which at present are more than 8 years, has been recommended to be reduced to 4 years. Application of 25 tons/ha animal manure is recommended. However, chemical fertilizers especially nitrogenous fertilizer (urea) at a rate of 100 kg/ha after flower harvesting has also been recommended by some workers. However, fertilizer recommendation should be based on soil analysis.

Although it is recommended that mechanical weed control be applied for a healthy product, for environmental concerns, when it is required, herbicides such as sencor at a rate of 750-1000 g/ha for broad leaves and gallant or sonalan at a rate of 1 to 1.5 liter/ha for narrow leaves could be applied. However production of organic saffron which has been carried out for centuries on the basis of traditional practices is still considered feasible and should be encouraged according to the International Certification Standards.

References

1. Abassi, M. A. 1997. Effects of herbicides on saffron weeds. M.Sc thesis, Faculty of Agriculture, Ferdowsi University of Mashhad, Iran.

2. Abrishami, M. H. 1988. Saffron of Iran. Toosi Publishing Company, Iran.

3. Ait-aubahou, A. and M. El-otamani. 1998. Saffron cultivation in Morocco. In: M. Negbi (ed), Saffron. Harwood Academic Pub. Amsterdam, Pp. 154

4. Alavi Shahri, H. 1996. Effects of rate of irrigation and animal manure on saffron yield. Journal of Plant and Soil. Vol. 11.

5. Alavi Shahri, H., M. Mahajeri and M. A. Falaki 1995. Effects of plant density on saffron yield. Abstracts of the Second National Symposium on Saffron and Medicinal Plants. 8-9 November 1994, Gonabad, Iran.

6. Amirghasemi, T. 2002. Saffron, red gold of Iran. Azadeghan Publishing Company.

7. Amiri, H. 1991. Effects of herbicides on weed control in saffron. Khorasan Agricultural Research Center (Technical report).

8. Azizi zehan, A. 2001. Saffron water requirements, methods of irrigation and irrigation intervals. M.Sc thesis, Faculty of Agriculture, Shiraz University, Iran.

9. Behnia, M. R. 1995. Effects of animal manure and chemical fertilizers on yield of saffron. Abstracts of the Second National Symposium on Saffron and Medicinal Plants. 8-9 November 1994, Gonabad, Iran.

10. Behnia, M. R. 1992. Saffron cultivation. Tehran University Press.

11. Eftekharzadeh Marghi, M. S. 1997. Evaluation of the effects of irrigation intervals and rate of N-fertilizer on flower production in saffron. Abstracts of 4th Iranian Crop Science Congress, Isfahan, Iran.

12. Farooq, S. and K. K. Koul. 1983. Changes in gibberllin-like activity in corms of saffron plant (*Corcus sativus* L.) during dormancy and sprouting. Journal of Plant Biochemistry. 178: 685-689.

13. Ghalavand, A. and M. Abdulahian. 1995. Effects of plant spacing and methods of planting on saffron. Abstracts of the Second National Symposium on Saffron and Medicinal Plants. 8-9 November 1994, Gonabad, Iran.

14. Goliaris, A. H. 1999. Saffron cultivation in Greece. In: Negbi. M. (Ed.). Saffron Harwood Academic Pub. Amsterdam, 154 pp.

15. Habibi, M. B. and A. Bagheri. 1989. Saffron: Cultivation, processing, chemical composition and standards. Iranian Scientific and Industrial Research Organization Khorasan Center (Technical report).

16. Halevy, A. H. 1985. Handbook of flowering. CRC Pub. Co. USA.

17. Herati, A. 1989. Effects of preplanting herbicide on saffron weeds. Abstracts of the Second National Symposium on Saffron and Medicinal Plants, Gonabad, Iran.

18. Housini, M. 1998. Effect of nutrient foliar application on yield of saffron. Iranian Scientific and Industrial Research Organization-Khorasan Center.

19. http://www.Iransaffron.org and www.Iraniansaffron.com

20. Kaith, D. S. and P. P. Sharma. 1984. Saffron in Himachal Pradesh: Retrospect and prospect. Indian Cocoa, Nuts and Species Journal. 7(1): 5.

21. Koocheki, A. 1996. Sustainable agriculture. Mashhad, Jahad Daneshgahi Publication.

22. Latifi, N. and K. Mashayekhi. 1997. Effects of corm size on saffron flower production. Abstracts of the Fourth Crop Science Congress, Isfahan, Iran.

23. Madam, C. L., B. M. Kapur and U. S. Gupta. 1966. Saffron. Econ. Bot. 20(4): 377-385.

24. Mashayekhi, K. and N. Lotfi. 1998. Effects of corm size on saffron flower production. Journal of Agriculture science. Vol. 28. No. 1.

25. Mehri, A. and M. R. Kahi. 2003. Design of a machine for separation of stigma from flower. Engineering Research Center of Jehad Keshavarzi, Khorasan (Technical report).

26. Modaresi, M. 1996. Saffron: a red gold. 1996. Proceedings of Khorasan Quality Control Society, Mashhad, Iran.

27. Molaphilabi, A. 2001. Production management in saffron. Iranian Industrial and Scientific Research Organization - Khorasan Center (Technical report).

28. Rage, M. A. Mobin and H. Faghih. 1989. Effect of herbicides on saffron weeds. Abstracts of the First National Saffron Symposium. 8-9 November 1994, Gonabad, Iran.

29. Raghimi, G.H. 1991. Machinery for saffron production. Faculty of Agriculture, Birjand University (Technical report).

30. Rahimian, M. 1994. Effects of herbicides on saffron weed control. Iranian Industrial and Scientific Research organization - Khorasan Center (Technical report).

31. Rashed Mohasel, M. H. 1989. Identification of saffron weeds in South Khorasan. Abstracts of the First National Symposium on Saffron and Medicinal Plants. 8-9 November 1994, Gonabad, Iran.

32. Rashed Mohasel, M. H. 1990. Reports of visit of scientific delegates to Spain on Saffron Iranian Scientific and Industrial Research Organization-Khorasan Center.

33. Reres Bueno, M. P. 1989. Elzafran. Ediciones, Mondi Prensa.

34. Rezaiyan, S. and M. Pasban. 2002. Effect of application of urea and sulfur coated urea on yield of saffron in South Khorasan. Khorasan Agricultural Research Center (Technical report).

35. Sadeghi, B. 1997. Effects of corm storage and planting date on saffron flower production. Iranian Industrial and Scientific Research Organization- Khorasan Center (Technical report).

36. Sadeghi, B. 1994. Effects of corm size on saffron flower production. Iranian Industrial and Scientific Research organization. Khorasan Center (Technical report).

37. Sadeghi, B., M. Razavi and M. Mahajeri. 1989. Effects of rate of fertilizer application on yield of saffron. Khorasan Agricultural Research Center (Technical report).

38. Sadeghi, B. 1988. Effects of nutrients on saffron production. Khorasan Agricultural Resarch Center (Technical report).

39. Saiidirad, M. H. 2002. Design of a corm planting machine. Khorasan Agricultural Research Center (Technical report).

40. Shahandeh, H. 1991. Evaluation of physical and chemical properties of soil and water associated with saffron yield in Gonabad. Iranian Industrial and Scientific Organization - Khorasan Center (Technical report).

41. Tammaro, F. 1999. Saffron in Italy. In: Negbi, M. (Ed.). Saffron. Harwood Academic Pub. Amsterdam, 154 pp.

42. Torabi, M. and B. Sadeghi. 1995. Pattern of nutrient changes in leaf and corm of saffron during growth period. Abstracts of the Second National Symposium on Saffron and Medicinal Plants. 8-9 November 1994, Gonabad, Iran.

43. Valizadeh, R. 1989. Evaluation of saffron leaf as animal feed. Iranian Industrial and Scientific Research Organization-Khorasan Center (Technical report).

CHAPTER 5

Irrigation

A. Alizadeh
College of Agriculture, Ferdowsi University of Mashhad, Iran

5-1 INTRODUCTION

Saffron is an ideal plant for arid and semi-arid regions with water
limitations because its corms have a 5-month dormancy period
without irrigation requirement, which starts from early May when
spring rainfalls are almost finished. Once out of its dormancy, saffron
have to be irrigated like any other crop. In the saffron producing
regions of Iran irrigation starts from mid-October to early November
depending on local climate. However, irrigation may start from early
October in cold regions and be delayed till late November in warmer
climates (1).

Growth of saffron starts immediately after the first irrigation and
flowering is the first stage of growth. Therefore, in regions with a
high cultivated area and limited labor, to avoid simultaneous
flowering, sequential irrigation of the farms is scheduled. In fact the
first irrigation is practiced when plants have not appeared on the soil
surface. Shortly after this irrigation flowers will appear and plant
development will follow later with leaf growth. The best time for the
second irrigation is about 4-5 weeks after the first. The next
irrigation scheduled with 12-14 days interval depending on water

availability and continues until May when leaf color changes to yellow. Irrigation usually stops after mid May (1, 4).

Based on the indigenous knowledge of Iranian saffron producers four irrigations should be enough for harvesting a good saffron yield.

1. The first irrigation is required for start of growth and facilitation of flowering. However, timing of this irrigation is crucial. If scheduled at a proper time, flowers will appear immediately after irrigation and vegetative growth will start later, otherwise flowering and vegetative growth will start simultaneously and the latter may interfere with harvesting practice.

2. The second irrigation is delayed until flowers are harvested and leaves appear. In practice it takes about a month after the first irrigation.

3. The best time for the third irrigation is after weeding and spreading fertilizers.

4. The last irrigation should be scheduled by the end of growing season (usually May).

Summer irrigation is not a common practice. However, Sadeghi (8) in a 2-year experiment showed that irrigation in July was harmful but irrigation in August led to an increase in saffron yield of both newly established and old saffron fields by 17 and 40%, respectively. Mosafer (6) also showed that irrigation in mid-June resulted in 17% reduction saffron yield but flower yield increased by 20% when summer irrigation was conducted in late-August. It has to be noted that summer irrigation will usually increase the risk of fungal disease.

Information about the water requirement of saffron is scarce. In fact in many countries such as Spain, parts of Italy and Kashmir saffron is produced in a rainfed system with no irrigation (7). In Spain where up to 90% of saffron is produced in a rainfed system, irrigation during August and September had positive effects on yield. However, it is usually avoided because of corm rot diseases. In Kashmir during the years with low summer rainfall, irrigation in September is a common practice to enhance flowering (7).

5-2 WATER REQUIREMENT

Saffron growth period corresponds with cold season and during the summer where water shortage is a limiting factor for growth of many crops, saffron stays dormant with no requirement for irrigation. However, obtaining high yields depends on enough water for vegetative growth. Therefore, information about consumptive use of this plant is crucial for planning proper irrigation regimes.

Like other crops consumptive use of saffron is calculated based on crop coefficient (K_c) and reference evapo-transpiration (ET_O). When ET_O is known, evapo-transpiration of saffron can be calculated from Equation (1):

$$ET_C = K_c (ET_O) \qquad ...(1)$$

Where ET_C is the consumptive water use of saffron. When only the pan evaporation data are available, ET_C will be obtained using Equation (2):

$$ET_C = K_p (E_p) \qquad ...(2)$$

Where K_p and E_p are pan coefficient and pan evaporation, respectively. Before application of Equation (2) pan coefficient for saffron should be obtained based on field experiments.

5-3 CROP COEFFICIENT

There is little information about water requirement of saffron. Lack of such applied research is mainly due to limitation in areas under saffron cultivation and the perennial nature of this plant. To determine water requirement of saffron lysimetric investigations have been conducted in Khorasan. Since 1998 during a series of experiments, consumptive water use of saffron compared to reference evapo-transpiration has been calculated in lysimeter and on this basis crop coefficients of saffron was determined (2; 5).

The values of saffron evapo-transpiration rate compared to reference and class A pan evaporation during growth period

(October-May) for 1998-1999 growing season are shown in Table 5-1. Based on these data monthly values of crop coefficient (K_c) and pan coefficient (K_p) were calculated for saffron. The highest K_c value of 0.75 was obtained during January-March period and pan coefficient reached its maximum in March (Table 5-1).

Table 5-1 *Reference evapo-transpiration (ET_O), crop evapo-transpiration (ET_C) and crop coefficients (K_c) of saffron for different months (1998-99). For a two years old farm field.*

Month	ET_O (mm day^{-1})	Class A pan evaporation (mm day^{-1})	ET_C (mm day^{-1})	K_c	K_p
Oct.	2.6	4.7	1.2	0.45	0.3
Nov.	1.8	3.3	0.9	0.50	0.3
Dec.	1.3	1.8	0.8	0.65	0.4
Jan.	1.0	2.8	0.7	0.75	0.4
Feb.	1.0	2.3	0.8	0.75	0.4
March	1.4	2.9	1.0	0.75	0.4
April	2.6	2.8	2.0	0.75	0.5
May	3.4	6.2	2.0	0.60	0.3

Saffron is a perennial crop and its vegetative cover increases due to corm propagation and canopy development. Therefore, its water requirement is different with the age of fields. Data in Table 5-1 were obtained from a 2-year old field. Water requirement of the same field during 1999-2000 growing season (3-year old field) are presented in Table 5-2. Results showed that in the third year of growth K_c increased by 16% compared to second year of growth. Observed changes in water requirements of saffron during the years are the result of canopy extension and more soil cover by older crop (2, 5).

Figure 5-1 shows calculated K_c values of saffron for two successive years. However, in practice crop coefficient curve is usually plotted by dividing plant growth period into four stages, assuming linear changes of K_c for each growth stage (3). Using data of Figure 5-1 four-stage K_c curve of saffron was generated as shown in Fig. 5-2.

Table 5-2 *Reference evapo-transpiration (ET_O), crop evapo-transpiration (ET_C) and crop coefficients (K_c) of saffron for different month (1999-2000). For a three years old field.*

Month	ET_O (mm day^{-1})	ET_C (mm day^{-1})	K_c
Oct.	2.5	1.3	0.52
Nov.	1.8	1.0	0.58
Dec.	1.0	0.75	0.75
Jan.	1.2	1.0	0.85
Feb.	1.3	1.2	0.85
March	1.3	1.8	0.85
April	3.8	3.0	0.80
May	4.7	3.2	0.68

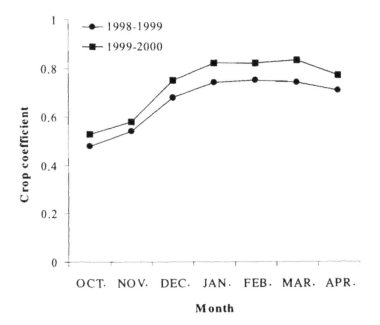

Fig. 5-1 Mean monthly crop coefficients of saffron for 1998-1999 and 1999-2000 growing seasons

Mean monthly values of evapo-transpiration/class A pan evaporation ratio during the growth period of saffron is shown in Fig. 5-3. This ratio ranges from 0.3-0.5 with a maximum in December and an average of 0.4.

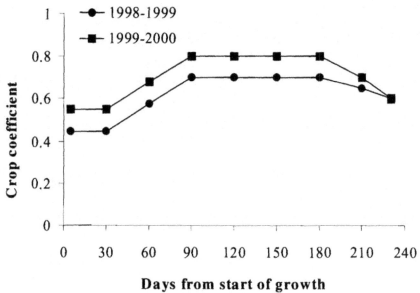

Fig. 5-2 Crop coefficient curve of saffron during different growth stages for 1998-1999 and 1999-2000 growing seasons (adapted from 7).

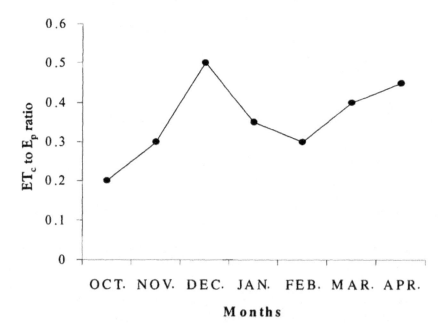

Fig. 5-3 Ratio of saffron evapo-transpiration (ET_c) to class A pan evaporation (E_p) during the saffron growth period

Table 5-3 *Water balance of lysimeter grown saffron during two successive seasons (1998-2000)*

Month	Irrigation number	Rainfall (mm)	Irrigation water (mm)	Drainage water (mm)	Evapotranspiration (mm)
DEC 1998	1	8	20	7.7	19.4
	2	-	20	7.0	13.9
JAN 1999	1	0	20	9.9	10.1
	2	-	20	5.9	14.1
FEB 1999	1	21.3	20	17.0	14.3
	2	-	20	15.0	14.9
MAR 1999	1	22.5	20	0.4	19.5
	2	-	20	7.5	25.1
APR 1999	1	3.5	40	7.7	35.3
	2	-	50	1.9	48.6
MAY 1999	1	0	40	7.7	32.3
	2	-	40	-	22.7
OCT 1999	1	0	20	5.5	14.5
NOV 1999	1	29	0	2.0	10.9
	2	-	0	2.0	13.1
DEC 1999	1	38.7	0	2.8	15.0
	2	-	0	5.0	15.0
JAN 2000	1	24	0	3.2	10.8
	2	-	10	6.3	13.6
FEB 2000	1	0	20	5.7	14.3
	2	-	25	7.3	17.7
MAR 2000	1	0	25	7.3	17.7
	2	-	20	8.0	35.0
APR 2000	1	6	35	4.0	37.0
	2	-	50	11.2	38.8
MAY 2000	1	3.7	40	8.0	35.7
Total	-	-	-	-	299.5

Water balance of saffron in a lysimeter experiment was measured during two successive years (6). Annual water requirement of saffron was estimated as 299.5 mm amounting to 3000 m^3 year^{-1} (Table 5-3).

Variation in monthly water requirement of saffron with maximum of 2.5 mm day^{-1} during April and May is shown in Figure 5-4.

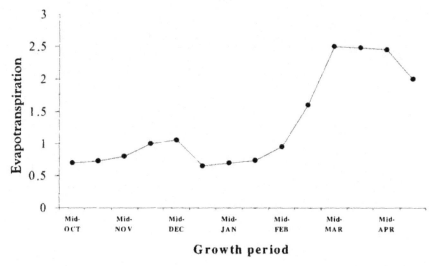

Fig. 5-4 Monthly variation in crop evapo-transpiration (mm day^{-1}) of saffron during growth period

K_c values of saffron reported by Mosaferi (6) were higher than that of Mahdavi (5). The lowest and the highest crop coefficient values of 0.41 and 0.98 respectively for October-November and February-March were reported by Mahdavi (5, Table 5-4).

Actual K_c curve based on the results of Mosaferi (6) and typical 4-stage K_c curve are shown in Figures 5-5 and 5-6, respectively. Duration of the four stages for saffron and their corresponding K_c values are given in Table 5-5.

Shirmohammadi (9) reported K_c values of 0.45, 0.80 and 0.30 for initial, mid-season and late season growth stages of saffron, respectively. Based on the results of these experiments for initial, mid

Table 5-4 *Monthly reference evapo-transpiration (ET$_O$), crop evapo-transpiration (ET$_C$) and crop coefficient values (K$_c$) for lysimeter grown saffron*

Month	ET$_O$ (mm)	ET$_C$ (mm)	K$_c$
October	62.1	42.5	0.41
November	26.2	10.9	0.41
December	41.8	28.1	0.75
January	31.2	26.7	0.88
February	28.6	27.9	0.98
March	47.0	39.9	0.98
April	83.7	72.0	0.86
May	125.5	74.5	0.59

and late season stages K$_c$ values of 0.4, 0.85, and 0.55 respectively, could be used as conservative values.

Irrigation Regime

There is limited information about effects of amount and intervals of irrigation on yield components of saffron. In this regard three aspects are of great importance:

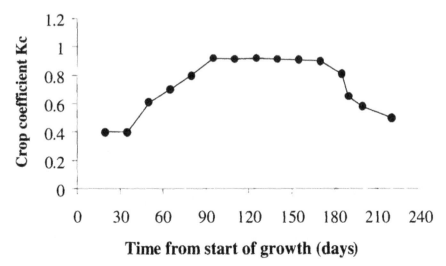

Fig. 5-5 Variation in crop coefficient of saffron during growing season

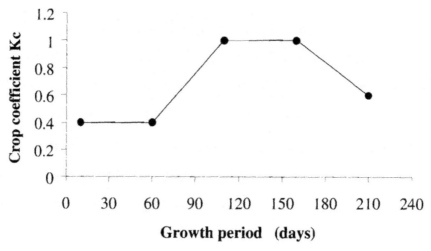

Fig. 5-6 Typical crop coefficient (K_c) curve for saffron.

- Effects of pre-flowering irrigation
- Effects after harvest irrigation on yield of the following year
- Effects of irrigation intervals on saffron yield.

In an experiment different dates and amounts of the first irrigation were investigated for their effects on saffron yield (6). Results showed that mid October with the highest fresh flower and dry saffron yield was the best time for the first irrigation (Table 5-6). Four levels of applied water with constant intervals of 15 days were used in each irrigation date. Irrigation levels were 10, 20, 40 and 80 lit m^{-2} corresponding to 100, 200, 400 and 800 m^3 ha^{-1}, respectively. Based on the results (Table 5-7) the highest saffron yield was obtained when 20 lit m^{-2} irrigation water was applied at 15 days intervals.

Table 5-5   *Duration of growth stages of saffron and corresponding K_c values.*

Growth stage	Duration (days)	Crop coefficient (K_c)
Initial	40	0.41
Development	55	-
Mid season	105	0.95
Late season	30	0.55

Table 5-6   *Effects of time of pre-flowering irrigation on saffron yield*

Time of first irrigation	Number of flowers m^{-2}	Flowers dry weight (g m^{-2})	Saffron yield (g m^{-2})
Early October	20.25	8.25	0.156
Mid October	30.50	11.02	0.226
Late October	23.75	8.47	0.170
Mid November	21.50	6.74	0.114

Table 5-7   *Effects of different levels of applied irrigation water on saffron yield*

Applied water (lit m^{-2})	Number of flowers (m^{-2})	Flowers dry weight (g m^{-2})	Saffron yield (g m^{-2})
10	18.5b*	25.01b	0.150b
20	25.7a	37.01a	0.160a
40	25. 0a	33.95ab	0.153b
80	21.25b	27.13b	0.117b

*In each column means with the same letters are not significantly different at P<0.05.

Table 5-8   *Effects of irrigation intervals on saffron yield*

Irrigation interval (days)	Number of flowers (m^{-2})	Flowers dry weight (g m^{-2})	Saffron yield (g m^{-2})
15	152a	12.46a	0.194a
30	126b	10.75ab	0.168ab
60	102b	3.95b	0.150b
Control (no irrigation)	101b	9.04b	0.138b

*In each column means with the same letters are not significantly different at P<0.05.

Effects of irrigation intervals (15, 30 and 60 days) were compared with no irrigation control (6). Results showed that all yield components of saffron were significantly higher in irrigation with 15 days' intervals with up to 50% increase in number of flowers compared to no irrigation control (Table 5-8).

These results are based on experiments conducted under climatic conditions of Mashhad, North East of Iran. More experiments in different climates and with saffron fields at different ages are required to schedule the proper irrigation regimes for saffron.

SUMMARY

Traditionally saffron is irrigated four times during the growing season. However, to achieve high yields sufficient water is required from October until May. Duration of initial, developmental, mid season and late season stages in saffron is 30, 55, 105 and 30 days, respectively and estimated crop coefficient values (K_c) for these successive stages are 0.4, 0.85 and 0.55. Under Korasan climatic conditions of mid October is the best time for the first irrigation. This should be continued with 15 days' intervals for maximum yield. Summer irrigation has positive effects on saffron yield. However, it is not recommended because of the high risk of fungal diseases.

References

1. Abrishami, M. H. 1997. Iranian saffron: historic, cultural and agronomic prospects: Astan Ghods Razavi Publishing Co. (Persian).
2. Alizadeh, A. 2001. Potential evapo-transpiration of cumin and saffron. Project Report. National Meteorological Organization.
3. Allen, R.G., L.S. Pereira, D. Raes and M. Smith, 1998. Crop evapotranspiration. FAO Irrigation and Drainage Paper No. 56. FAO, Rome, Italy.
4. Habibi, M.B. and A.R. Bagheri, 1989. Saffron, agronomy, processing, chemistry and standards. Scientific and Industrial Research Organization of Iran, Khorasan Center.
5. Mahdavi, M. 1999. Crop coefficient and evapo-transpiration of saffron under standard conditions. MSc. thesis, Ferdowsi University of Mashhad, Iran.
6. Mosaferi Ziaedini, H. 2001. Effects of different irrigation regimes on saffron yield. MSc. thesis, Ferdowsi University of Mashhad, Iran.
7. Poglini, M. and F.D. Groose, 1971. Stadi cariologico di *Crocus sativus*. Inform. Bot. Ital. 4: 25-29.
8. Sadeghi, B. 1998. Effects of summer irrigation on saffron yield. Scientific and Industrial Research Organization of Iran, Khorasan Center.
9. Shir Mohammadi, Z. 2002. Effects of method and amount of irrigation water of lean area index, canopy temperature and yield of saffron. MSc. thesis, Shiraz University, Iran.

CHAPTER **6**

Saffron Pests, Diseases, and Weeds

M.B. Shahrokhi[1], H. Rahimi[1] and M.H. Rashed[2]
[1]Khorasan Agricultural Jahad Research Center,
[2]Ferdowsi University of Mashhad

6-1 INTRODUCTION

Considering the importance of saffron and its role in Iran, Kashmir in India and many other countries improving the agriculture of southern and central great Khorasan, advancement programs in qualitative and quantitative production of this valuable crop requires expanding efforts. One of the main problems in saffron production is the presence of relatively diverse pests including rodents, insects, and plant mites; diseases such as corm rot; and weeds. Some of the species have a wide range of activities and cause damage to saffron. If rational control measures are not considered, it will result in damage such as destruction of farm and yield reduction.

During the last few decades, the Plant Pests and Diseases Research Department in Agricultural Research Center of Khorasan have done studies and experiments concerning the harmful agents of saffron. The main pests that have been identified for their ability to damage saffron will be discussed here.

6-2 PORCUPINE *(HYSTRIX INDICA)* KERR

Porcupines are the largest rodent that attack saffron plantations. The length of its body reaches up to 80 cm. The distinctive features of a porcupine are a heavy stout body, variegated brown color, with long and sharp quills, which are, in fact, a defensive mechanism against enemies. Porcupine nests are located inside mountain gaps, wells or in destroyed kanats.

Saffron is one of the porcupines favorite foods. They love saffron corms. Their nests may be far away from saffron fields, but the presence of corm sheath, and disturbed soil around the plants and across the field are indicators of porcupine activity. We usually observe more damage on hill side fields which are usually close to porcupine nests. In order to prevent severe damage, the following methods are effective.

1. application of toxic gas inside the nest
2. using toxic bait inside the fields
3. building barriers
4. predation.

6-3 BANDICOOT RAT (*NESOKIA INDICA* GRAY)

Bandicoot rats are about 14 to 20 cm long. A lepidopted tail with no hair is about 2/3 the size of the body. The body is gray, light brown and dark brown. The upper front teeth are strong with a width more than the nasal bone. Bandicoot rat nests are branched and composed of 5 to 15 holes with a mass of soil around them. The depth of the nest is about 95 cm.

Bandicoot rats are active in most saffron plantation areas. Since this rat eats saffron roots and corms, they attack plants within the soil. Bandicoot rats activityis seen in digging and destruction of the soil, water runoff, and destroyed plants. Bandicoot rats are active throughout the year and produce 5 to 6 generations. Females give birth to 3-9 babies each time.

Control measures:

1. Using toxic zinc phosphide bait with grass or wheat.
2. Using kelerat
3. Using rat traps
4. Cultural practices.

6-4 MOUSE (*MUS MUSCULUS* L.)

Mice are the smallest saffron field rodent, about 6-9.9 cm long with a tail about the same size as its body. The overall color of mice is light brown and gray. Although mice are present within saffron fields, they flourish on grass and saffron leaves. So far, large populations of mice are not observed in saffron covered areas.

6-5 AFGHAN MOLE RAT (*ELLOBIUS FUCOCABILUS* BLYTH)

The Afghan mole rat is a relatively small to medium size rodent, 11-14 cm long with a small tail. Its tiny eyes are usually covered under hair and that is the reason that Afghan mole rats have a relatively low vision and are mainly active inside their nests. The back of this rat is light brown while the head is dark brown. Afghan mole rat nests have many branches and are extremely labyrinthous. The nest opening is covered under soil heaps as a result of digging holes, and it shows the activity of blind rats within saffron fields. The holes are 15-25 cm deep.

Afghan mole rats are mostly active under the ground and attack plant roots and corms. In saffron fields Afghan mole rats prefer corms and flourish on it. Afghan mole rats rarely exit from the nest and eat grass. They are active throughout the year and live mostly on hillsides. Therefore, in newly formed saffron fields in mountainous areas and foothills, the density of Afghan mole rats is remarkable.

In some areas of Ghaen such as Abeez, the activity and damage by Afghan mole rats is high and this rat is considered the main pest in much of these areas. The harmful effects of Afghan mole rats are digging the soil and making labyrinth types of nests with several holes resulting in destruction of the fields. The best method of Afghan mole rat control is using fumigant toxin especially phostoxin. In these methods placing 1-2 peels within 1 hole while closing the other holes results in volatility of PH3 gas which kills the rats. In order to produce more gas the relative humidity should be high. During the summer when the soil is dry we should spray water within the nest. The practical way is putting peel within a wet cloth and placing it inside the holes.

6-6 INSECTS

In a study by Rahimi (7) on identifying and introducing insect pests and their natural enemies in south Khorasan saffron fields, he presents a list of harmful and useful insects (Table 6-1). Among 16 species cited above, Saffron bulb mite (*Rhizoglyphus robini*) and corm thrips (*Thrips tabaci*) have the ability to cause economic damage to saffron plantations. These two species will be discussed in more detail and some of the others will be mentioned briefly.

6-6-1 Saffron Bulb Mite (*Rhizoglyphus robini*)

This is a cosmopolitan species, which is one of the most important corm pests. Manson (5) have mentioned garlic, carrot, gladiolus, species of iris, narcissus, dahlia, onion, potato, amaryllis and tuber plants act as hosts of this mite. Rahimi reported this mite from saffron fields in Gonabad and Ghaen (7). In 2002 this mite was reported from saffron corms in most saffron fields in central and southern Khorasan. However, infestation is more severe in areas with a long history of saffron growing or areas with lack of proper sanitation.

Morphology

The mite is about 0.6-0.8 mm long with a bulging oval body, dark appearance, and slow mobility (Figure 6-2). The legs of this bulb

Table 6-1 *Useful and harmful organisms collected from saffron fields of Southern Khorasan (7)*

Scientific name	Family	Host	Active time's	Importance
Rhizoglyphus robini	Acaridae	Corm	All times	High
Tyrophagus putrescentiae	Acaridae	Corm	All times	Very low
Petrobia latens	Tetranychide	Leaf	Fall to early Spring	Very low
Bryobia praetiosa	Tetranychide	Leaf	Fall to early Spring	Very low
Penhtaleus major	Penthaleidae	Leaf	Fall to early Spring	Very low
Thrips tabaci	Thripidae	Leaf	Fall to early Spring	Moderate
Collemblothrips Sp.	Thripidae	Leaf	Fall to early Spring	Very low
Haplothrips reuteri	Phlaeothripidae early Spring	Leaf and Flower	Fall to early Spring	Very low
Myzus certus	Aphididae	Leaf	Early Spring	Very low
Aulacorthum palustre	Aphididae	Leaf	Early Spring	Very low
Xenylla boerneri	Hypogastruridae	Corm fibers	All times	Very low
Sphaeridia Pumilis	Sminturidae	Corm fibers	Spring times	Very low
Hippodamia variegate	Coccinellidae	Aphid	Early Spring	Low
Coccinella septempuntata	Coccinellidae	Aphid	Early Spring	Very low
C. undecempuncata	Coccinellidae	Aphid	Early spring	Very low
Oenopia conglobata	Coccinellidae	Aphid	Early Spring	Very low

mite are short and stiff, reddish brown with numerous hair. In nature we see different forms including female, non-similar and similar appearance to male, and Hypoppus.

Biology

The eggs are oval, initially clear but gradually turn dirty white. About 24 hours before hatching a red spot appears on one side of the

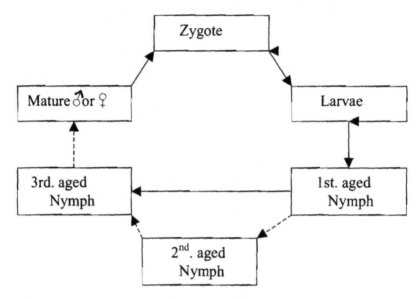

Fig. 6-1 Diagram showing the life cycle of saffron corm mite.

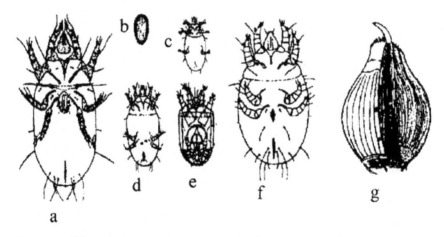

Fig. 6-2 Different life cycle stages of *R. robini* (a= mature mite. b= zygote, c= larvae, d= 1st. aged nymph, e= 2nd. aged nymph, f= 3rd. aged nymph); and g= damaged corm.

egg. At this time legs and red mouth appendage can be observed easily within the egg. Rahimi (6) reported the length of embryonic growth and development of saffron bulb mite at $25 \pm 1°C$ on saffron corm is between 4-5 days with an average of 4.25 days.

Larvae

Young larvae coming out of hatched eggs are clear with three pairs of reddish brown legs like most mites. Flourished mite turn into non-transparent white and become stagnant and enter the first resting nymph chrysalis. Rahimi (6) mentions the larvae stage with duration of about 2 to 3.5 days and the resting stage about 0.5 to 1 day and 0.81 day average.

After the first resting stage the 4th pair of legs appears and the larvae goes into nymph period. Bulb mites have 3 nymph ages (Figure 6-1) and then turn into adult mite. On an average, the life cycle duration of this pest from egg to adult on saffron corm at 25 ± 1°C is about 13.75 days and 15 generations per year. Rahimi (6) at 25 ± 1°C have identified the life cycle of male on saffron corm is 60 to 115 days with a mean of 80 days and female life cycle about 40 to 80 days and a mean of 60 days. The life history of R. *robini* is shown in Figure 6-1.

Variation within population and generations

Saffron bulb mite is active throughout the year and is able to produce several generations. However, we see the maximum numbers in population and generations during spring and fall when optimum condition exists for growth and development. During summer, because of heat and dryness of soil and in winter, due to cold and freezing, the population reduces considerably. Irrigation in summer season when saffron is in resting stage will result in providing suitable condition for growth of corm of saffron bulb mite and a rapid increase in the numbers.

Method of damage

Bulb mite attacks saffron corms through wounds of sometimes healthy parts of the corm. By producing bores and tunnels and cavity in corms the mite starts reproduction within this cavity. The cavity expands gradually and putrified substances penetrate from the wounded areas and cavity within the corm and accelerate putrification.

The plants infested with mites have short and slender leaves. The infested leaves die earlier than healthy ones, and the gradual damage of this pest is thinning saffron fields.

Prevention and control of saffron bulb mites are:

a) In established fields

The remarkable activity of mite observed within the fields in which corms are shallower than optimum depth, or in fields being irrigated during summer or cracked corms have been planted, so we may recommend:

1. Do not irrigate saffron during the summer
2. In fields with shallow corms, add light soils up to 15 cm depth during summer when corms are in resting stage. In such cases corms are not exposed to serious damage.
3. The first irrigation (before flowering) and the second irrigation (after flowering) should be done adequately and carefully
4. Weed control must be done with care.

b) In new fields

1. Pull up saffron corms for planting in new fields and avoid irrigation to facilitate the action
2. Choose healthy and uniform corms for planting
3. Treat the corms with fungicide – miticide before planting
4. The depth of planting should be 15 to 20 cm depending on soil texture
5. During summer of every other year add 1-2 cm light soil to the ground in order to keep depth of planting constant
6. Avoid irrigation of saffron fields during summer because the most important factor of mite activity, soil moisture, becomes available to mite and increases the mite activity and population

7. Do not remove the superficial soil for saffron reseeding
8. Use high rate of corms per unit area to reach economical harvesting levels and mites do not have enough opportunity for population growth
9. Use rotten seedless organic cow manure. Farmers' experience indicates that sheep and chicken manure are not suitable for saffron
10. Do not transfer contaminated corm to other areas.

6-6-2   Corm Thrips (*Thrips tabaci* Linedman)

A worldwide insect and polyphagus which is active on saffron leaves from beginning of saffron growth up to the point they die. The peak activity period is from early March to late April.

Morphology

The adult is about 0.8-1 mm long, yellow to light brown; wing's narrow and sharp, pointed with long hair at the rear and 2 longitudinal veins on front wings. Pronotum without long lateral hair and rear margin with 4 short hair about 48 μ; head wide, length less than width with 3 simple eyes and a pair of compound eyes; antenna with 7 joints, the first joints lighter than others in color; mouth parts rasping- seeking type, the upper and lower mandible stylet shaped and elongated into a trunk. Thrips Nymph yellow with lighter color than adult.

Biology

Adult insect in winter on the leaves of host plants such as saffron, alfalfa, and weeds or on the soil surface and plant debris. After flourishing for 20-25 days, the female lays eggs by early March. Nourishment is done by rupturing plant epidermis, scraping tissues, penetrating stylet into plants tissues, and sucking plant sap. The results not only weaken and inhibit plant growth but also transform viral pathogens from infected plants to healthy plants. Female thrips nurture more than male thrips and may reproduce sexually or parthenogenesis

The female lays 80-100 eggs within plant tissues. The eggs hatch after 3-4 days and the nymph is mostly located within two grooves on the abaxial side of saffron leaves. Nymph stage is between 9 and 12 days before they stay underground as pupae and stay there until maturity. The life cycle of thrips from egg to maturity is 20-22 days. This multigenerational thrips is able to spend 1-2 generations on saffron and other generations on different host plants.

Method of damage

The presence of yellow to white spots on saffron leaves are an indication of thrips damage. The severely infected leaves senesce earlier and result directly in saffron yield reduction.

Meanwhile, it has been proven that this thrips transfers several viral diseases in different plants, since saffron reproduces asexually. The possibility of virus or viral diseases needs further investigation.

Prevention and control

Since a high population of this pest is observed at the end of the saffron growing season, chemical control is not necessary. However, if the population increase is distressing, chemical control measures should be taken.

6-7 SAFFRON DISEASE

Jafarpour (4) after preliminary investigation reported the most important saffron diseases as follows:

6-7-1 Stigma twisting

In this disease the stigma twists like a spring, and in some cases the frilled stigma falls off (Figure 6-3). The reason is unknown, but

Fig. 6-3 Stigma twisting of saffron

possibilities such as the age of corms, plant nutrition and soil type, plant physiological conditions, and presence of plant viral and mycoplasmatic agent's type, which needs more investigation, should also be considered.

6-7-2 Corm rots

Corm rots may results from fungi such as *Rhizopus*, *Aspergillus*, and *Penicillium* which are common in stored corms (Figure 6-4). The common rots observed in Khorasan are as follows:

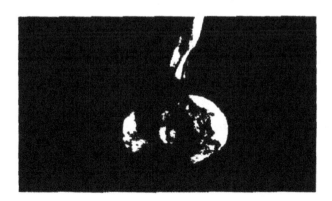

Fig. 6-4 Saffron corm rot due to soil borne fungi attack

6-7-2-1   Corm root rot

This disease was observed in Torbatehaydarieh saffron fields. The intact plants and corms are smaller than normal. The reason is due to consuming a considerable amount of energy to develop spindle shape tubers under the corms. Possibilities such as improper conditions of irrigation, soil fertility and other improper conditions of soil may be the basis for this disease; presently this disease is not of economic relevance.

6-7-2-2   Corm neck rot

This infection is rarely observed within the fields and does not have economic importance.

The *Rhizoctonia crocorum*, which causes saffron death has been reported from Spain and France. The disease agent attacks external sheaths of the corm and damages the corms (8). The symptoms of root rot are yellowish leaves and presence of white mold spots on the corm. As the disease advances, the molds turn violet and cover the inner and outer portion of the corms. In most advanced cases pale red or reddish violet wrinkles are observed on the corm and the corm eventually rots. Sometimes the agent of corm rot may be *Rhizoctonia violacea* (10). In order to prevent this fungal disease, it is necessary to select healthy corms, and separate one or two layers of corm sheath before planting. To sterilize, embed the corms within 5% copper sulphate. When contamination is high we have to gather all the corms from the soil and destroy them (8). Crop rotation, sterilizing the soil twice with carbon sulfide, and sterilizing corms is recommended against these fungal diseases (2).

6-7-2-3   Saffron smut (Tacon)

This agent is a fungus named fumago, which grows on leaves and corms. The control measure is burning the infested leaves and corms (3).

6-7-2-4 *Saffron leaf chlorosis*

Leaf chlorosis may occur due to poor nutritive elements or liming of the soil or iron deficiency. Adding alkaline fertilizer and spraying the field with 2/1000 to 3/1000 sulphate solution (iron fertilizer) is effective for this physiological disease(3).

6-8 SAFFRON WEEDS AND WEEDS CONTROL

Saffron is a perennial crop, and since it is a short plant with upright narrow leaves, it is not a competitive plant and we see a variety of different species in saffron plantations. Weed control is an important step in promoting the quantity and quality of saffron. The first step in saffron weed control is identification of the weeds. Wrong identification of weeds results in wrong weed control programs. Rashed (9) performed a complete survey in several sites around Birjand, Ghaen, and Gonabad cities and he found 184 species within 128 genera and 32 families, from which, 113 species were annual and 71 species were perennials (including biennials). Asteraceae, Fabaceae, Poaceae, Brassicaceae and Chenopodiaceae respectively had the highest number of species (Table 6.2).

Table 6-2 *The number of weeds genera and species within distinct families in southern Khorasan (Ghaen, Gonabad, and Birjand) saffron fields (9)*

Family	Number of	
	Genera	*Species*
Asteraceae	25	32
Fabaceae	12	24
Poaceae	15	21
Brassicacea	14	19
Chenopodiaceae	6	12
Boraginaceae	8	10
Caryophyllaceae	7	10
Ranunculaceae	4	7

Since saffron is a perennial crop and fall flowering, the fall weeds are usually late flowering summer weeds, or winter rosette weeds that do not grow remarkably during fall and winter. Rashed concluded that early spring evaluation is the best criteria for winter weeds. Late spring evaluation will include summer weeds. Thus he did his evaluations from 10th-15th April (early spring) and from May 28-June 5 (late spring or early summer) within fixed 1m × 1m microplots. The results are shown on Table 6-3. The results of 10 dominant species during spring and summer evaluations in Gonabad, Ghaen, Birjand and throughout these regions are also shown on Table 6-4.

Table 6-3 *The average weed number and weed biomass dry weight (grams/m²) in Ghaen, Gonabad, and Birjand saffron fields during spring and summer evaluations (9)*

City	Number of weeds (plants/m²)		Weeds biomass (grams/m²)	
	Spring	Summer	Spring	Summer
Ghaen	316	109	286	176
Gonabad	300	161	354	333
Birjand	214	91	248	489

For instance, Table 6-4 shows that field bindweed (*Convolvulus arvensis*) is ranked 3 in the second evaluation and overall elevation in Ghaen and Birjand and hoary cress (*Cardaria draba*) in Gonabad is ranked 1 throughout the season. A common aspect observed in saffron fields states the fact that by progressing from spring to summer, winter annuals and perennials are replaced by summer annuals and especially summer perennial plants. This characteristic is mostly attributed to lack of saffron greens during the summer, fall flowering and perennial nature of saffron.

Based on observations and evaluations, weeds in southern Khorasan saffron plantations are a serious problem. Except for hand weeding, no other technique is being used and this requires tremendous amounts of energy. We have to look for another low cost method of control. Diversity of weeds, some of them with the same

Table 6-4 The ten most important saffron field weeds are ranked in cities of Ghaen, Gonabad, and Birjand and throughout the region. Evaluation is done during spring, summer, and total. 1 = most important, 10 = least important (9, 11)

Overall evaluation	Summer evaluation	Spring evaluation	Rank	Area
Hordeum pontaneum	Polygonum aviculare	Hordeum spontneum	1	
Polygonum aviculare	Acroptilon repense	Bromus tectorum	2	
Convolvulus arvensis	Convolvulus arvensis	Poa bulbosa	3	
Bromus tectorum	Alhagi persarum	Cardaria draba	4	
Cardaria draba	Ceratocarpus arenarius	Achillea wilhelmsii	5	Ghaen
Acroptilon repense	Chenopodium murale	Allium rubellum	6	
Poa bulbosa	Plantago lanceolata	Convolvulus arvensis	7	
Achillea wilhelmsii	Cardaria draba	Stellaria kotschyana	8	
Alhagi persarum	Lactuca seriola	Veronica campylopoda	9	
Ceratocarpus arenarius	Descurainia sophia	Holosteum glutinosum	10	
Cardaria draba	Cardaria draba	Cardaria draba	1	
Hordeum spontaneum	Hordeum spontaneum	Hordeum spontaneum	2	
Achillea wilhelmsii	Achillea wilhelmsii	Lamium amplexicaule	3	
Lamium amplexicaule	Centaurea bruguierana	Holosteum glutinosum	4	
Veronica ampylopoda	Polygonum aviculare	Poa bulbosa	5	Gonabad
Phalaris minor	Phalaris minor	Achillea wilhelmsii	6	
Holosteum glutinosum	Veronica campylopoda	Veronica campylopoda	7	
Centaurea ruguierana	Lolium rigidum	Erysimum barbaratum	8	
Descurainia sophia	Descurainia sophia	Malcolmia africana	9	
Polygonum aviculare	Lamium amplexicaule	Descurainia sophia	10	

(Table 6.4 Contd.)

(Table 6.4 Contd.)

		Birjand		Overall region
Bromus tectorum	*Alhagi persarum*	*Bromus tectorum*	1	
Cardaria draba	*Lactuca serriola*	*Holosteum glutinosum*	2	
Convolvulus arvensis	*Convolvulus arvensis*	*Cardaria draba*	3	
Alhagi persarum	*Polygonum aviculare*	*Hordeum spontaneum*	4	
Lactuca serriola	*Cardaria draba*	*Poa bulbosa*	5	
Descurainia sophia	*Plantago lanceolata*	*Descurainia Sophia*	6	
Polygonum aviculare	*Acroptilon repense*	*Lamium amplexicaule*	7	
Holosteum glutinosum	*Chenopodium murale*	*Convolvulus arvensis*	8	
Hordeum spontaneum	*Achillea wilhelmsii*	*Veronica campylopoda*	9	
Lamium amplexicaule	*Descurainia sophia*	*Allium altissisimum*	10	
Cardaria draba	*Cardaria draba*	*Cardaria draba*	1	
Hordeum spontaneum	*Hordeum spontaneum*	*Hordeum spontaneum*	2	
Achillea wilhelmsii	*Polygonum aviculare*	*Poa bulbosa*	3	
Convolvulus arvensis	*Alhagi persarum*	*Holosteum glutinosum*	4	
Polygonum aviculare	*Convolvulus arvensis*	*Bromus tectorum*	5	
Poa bulbosa	*Achillea wilhelmsii*	*Lamium amplexicaule*	6	
Lamium amplexicaule	*Lactuca serriola*	*Achillea wilhelmsii*	7	
Bromus tectorum	*Acroptilon repense*	*Veronica campylopoda*	8	
Veronica campylopoda	*Centaurea bruguierana*	*Descurainia Sophia*	9	
Hordeum glutinosum	*Descurainia sophia*	*Erysimum barbaratum*	10	

degree of competition, may sometimes result in creation of another problem resulting in dominance of another weed species. For example, two-rowed barley *(Hordeum spontaneum)*, Downy brome *(Bromus tectorum)*, and hoary cress *(Cardaria draba)*, 3 competitive species, was found in 3 areas under study and if we concentrate on hoary cress control we have to watch for downy brome and for two rowed barley. Microclima and time of evaluation have an important role in weed diversity. Research indicates that winter annuals have been gradually replaced by summer annuals and perennials (11). Weeds such as *Holosteum glutinosum*, Downy brome, and bulbous bluegrass *(Poa bulbosa)* appear in early spring while camel thorn Lambsquarter *(Chenopodium album)*, Russian thistle *(Salasola kali)*, knot weed *(Polygonum aviculare)*, and prickly lettuce *(Lactuca serriola)* appear in late spring and summer.

Saffron weed control in Iran is usually done by hand weeding. However, Abbasi (1) have conducted a series of experiments on weed control by using herbicides. His results indicate that metribuzine, when applied post emergence (0.7 kg.ai./ha) or pre-emergence (0.7 kg.ai./ha), also Fluazifop-p-butyl (0.25 lit.ai./ha), Haloxyfop ethothyle (0.25 lit.ai./ha) plus Metribuzine (0.7 kg.ai./ha), atrazine (0.8 kg.ai./ha), and ethalfluralin (1.16lit.ai./ha.) controlled saffron weeds better than other treatments. However, a minimum amount of weeds were observed in metribuzine pre-emergence or post-emergence, and mixture of haloxyfop etothyl plus metribuzine. Metribuzine, haloxyfop, fluazifop, and atrazine controlled graminaceous weeds and had a pronounced effect on wild barley. Bentazon (1.92 lit.ai/ha) had a remarkable effect on broad leaves but not narrow leaves. Combination of bentazon (1.92 lit.ai./ha) and atrazine (0.8 kg.ai./ha) controlled tansy mustard effectively. However, ethalfluralin and fluazifop could not effect broad leaves.

Post-emergence application of haloxyfop, fluazifop, and metribuzine and pre-emergence application of ethalfluralin did not have harmful effects on saffron flowering.

In conclusion, it is noteworthy that researches on weed control in saffron are not enough and we need a thorough investigation concerning weed identification and control in saffron fields.

SUMMARY

Saffron spends most of its growth period in winter at low temperature. Thus it faces low insect pests, instead, rodents are the main saffron pests. Porcupine, Afghan mole rat, bandicoot rat are among the main rodents, which attack saffron fields. The most important methods to control these pests is using poisonous baits and trap. Saffron corm mite is also reported from corms and leaves of saffron throughout central and southern Khorasan. This mite is active all year long, but it has a high population growth during fall and spring. Saffron corm mite can be controlled by cultural practices, mainly weed control and no irrigation during summer season. Thrips is a pest that is usually nurtured from the sap of saffron leaves and may transmit viral diseases.

Data concerning saffron diseases is not sufficient. However, saffron diseases are not a serious problem in Iran. So far, introductory studies have been done about corm rot, corm trunk rot, root rot, leaf chlorosis, and saffron smut (Tacon). The biology and life cycle of these agents and control measures have not been studied.

The most important weeds in saffron are *Cardaria draba* and *Hordeum spontaneum*. Winter annuals were mostly present during spring. However, as the season progressed, they were gradually replaced by summer annuals and some perennials. Weed control is by traditional hand weeding or hoeing but recent studies with chemicals indicates that herbicides such as metribuzin, haloxyfop, fluazifop, and ethalfluralin are promising and should be considered in future investigations.

References

1. Abbasi, E. 1997. The effects of herbicides on the weeds of *C. sativus* fields. M. Sc. Thesis, Faculty of Agriculture, Ferdowsi University of Mashhad, Iran.
2. Abrishami, M. H. 1988. Saffron of Iran. Toosi Publishing Company, Iran.

3. Behnia, M. R. 1988. Saffron, cultivation, management, harvesting, and marketing. Educational hand out, University of Birjand.

4. Jafar pour, B. 1988. An introductory investigation on saffron diseases. Iranian Industrial and Scientific Research Organization, Khorasan Center, (technical report).

5. Manson, D. C. M. 1972. A contribution to the study of genus *Rhizoglyphus claparede* 1869 (*Acarina, Acaridae*) Acarlogia. T. XÉÉÉ, Fasc. 4: 621 – 650.

6. Rahimi, H. 1994. Biological laboratory study of *Rhizoglyphus robini* mite and its damage to saffron corms in Ghaen and Gonabad. Journal of Agricultural Science. Shahid Chamran University of Ahwaz Publication. Ahwaz, Iran.

7. Rahimi, H. 2003. Identification and introducing insect pest and their natural enemies in saffron plantations of southern Khorasan areas. Khorasan Agricultural Research Center Publications.

8. Rahimi, H., and A. Bagheri. 1990. Saffron (cultivation, processing, chemical composition, and its standards). Iran Scientific and Industrial Research Organization. Khorasan Center.

9. Rashed-Mohassel, M. H. 1993. Weeds of south Khorasan saffron fields. Journal of Agricultural Science. Faculty of Agriculture, Ferdowsi University of Mashhad, Mashhad, Iran.

10. Rashed-Mohassel, M. H., A. Bagheri, B. Sadeghi, and A. Hemmati. 1990. Reports of visit of Scientific deligates to Spain on Saffron Iranian Science and Industrial Research Organization. Khorasan Center.

11. Reshinger, K. 1964. Flora Iranica. Academische Druck-U. Verganstalt, Graz, Austria.

CHAPTER 7

Genetics, Sterility, Propagation and *in vitro* Production of Secondary Metabolites

A. Bagheri and S.R. Vesal
College of Agriculture, Ferdowsi University of Mashhad, Iran

7-1 KARYOLOGY

Saffron genome contains 24 chromosomes that are morphologically separated in three groups of eight. Its karyotype structure shows little diversity within and between Asian and European populations (7). Using flow cytometry method, genomic size in saffron (C. *sativus* L.) was studied and small differences were detected (6). However, heteromorphism is observed in Majorka, Spanish (7) and La Acoila and Italian (10) populations. In the first two populations heteromorphism was detected in two chromosomes but in the other two populations it was observed only in one chromosome. Conjunction of trivalent chromosomes during meiosis in Japanese (54), Italian (10) and Iranian (41) populations confirm autotriploidy of this species. Based on studies on Italian populations, mean frequency of trivalents in the first metaphase was 7.3 per cell.

Meiotic studies in Iranian saffron have shown the presence of 8 trivalents during the first metaphase in 85% of pollen cells.

Therefore, Iranian saffron is also triploid with 8 trivalent (1, 34). However, trivalent frequency in natural populations of Kashmir saffron is about 64% and show autotriploidy in these populations (29). Triploidy of saffron leads to production of asymmetric gametes with frequency of up to 99%. These gametes are the main cause of sterility in this species (34). Gametes with 8 and 16 chromosomes are usually balanced and fertile, but due to lagging and inversion of chromosomes during the first meiotic anaphase, such gametes were not detected in saffron (35).

Different studies confirmed clear ploidy disorders during meiosis. During the first meiosis, irregular separation of chromosomes occurs due to formation of trivalents. Therefore, one or two chromosomes of each trivalent could be moved to spindle poles. This results in unbalanced distribution with 8 to 15 chromosomes in each pole. Unbalanced distribution could also be due to single chromosomes that cannot reach the poles or other disorders during chromatid division. Lack of genetic balance of nuclei will increase in the second meiosis as a result of laggards and extra polar assortment of chromatides during the second anaphase. Because of abnormal assortment of chromosomes, gametes are different from the tetrads that form in a normal meiosis (10, 41). (Figure 7-1)

7-2 DIVERSITY AND MUTAGENESIS

Despite distribution of more than 100 species of *Crocus* over Asia and Europe, ornamental species are of more interest and extensive genetic engineering is conducted for their improvement. However, triploid genome of cultivated saffron (*Crocus sativus*) is the main barrier of breeding programs (35).

Greece and Turkey are known as the centers of saffron diversity. In Iran, in addition to cultivated saffron, 8 wild species are reported. Among them *C. alemehensis*; *C. michelsoii* and *C. gilanicus* are native (35).

Fig. 7-1 Longitudal section of saffron anther with surrounding cells (T) and microspores after releasing from callus (x 259).

Cloning is the most practical method for saffron improvement. During the centuries of vegetative propagation, saffron has been subjected to many mutations. In fact, saffron is a mixed population consisting of many different colons. Therefore, isolation and identification of these colons for introducing new cultivations are of great importance (2).

Within the floral parts of saffron, stigma and its branching pattern are commercially important. However, little morphological diversity is reported in the flower structure of saffron. In normal flowers, stigma has three braches but stigmas with 3-10 branches are rarely reported. It seems that increase in stigma branches could be the result of combination of several flowering buds. Genetic studies have shown that all of these abnormal branches have 24 chromosomes. Therefore, there is no genetic basis for repetition of this trait (5, 1).

Increasing genetic diversity using induced mutations by gamma radiation or colchicine was studied in Iran. To increase polyploidy and chromosome enhancement, saffron corm buds were subjected to 0.1% colchicine treatment (5). In this experiment 48-58% of treated plants died and the remaining had a poor root system. Mitosis studies on meristems of initial leaves showed the same chromosome number as normal plants. Experiments with different doses of gamma radiation were not successful in induction on mutation. However, threshold of radiation tolerance in saffron was detected as 750-1250 Rad. It seems that there is no possibility for chromosome enhancement in saffron.

7-3 MIROSPOROGENESIS AND POLLEN DEVELOPMENT

Within the pollen sac, pollen mother cells are surrounded with four cell layers including loculi, central cell, tepetum and epidermis. During the meiosis pollen mother cells separate from each other and their shape will changed from polyhedral to spherical. Later, a callus layer will form on their external surface. Some pollen mother cells are subjected to cytogenetic changes such as cell deformation or cytoplasmic decay (8, 12). These changes that occur after formation of callus layer could be observed in one or more pollen mother cells and influence many microspore mother cells. This results in failure of microspores development together with abnormalities in loculi (12). These secretory tissues destroy and enter the pollen sac and produce a syncytiam layer around the microspores (69). Microspores in different sizes are released after collapse of callus layer (Figure 7-2).

Cytogenetic and structural development of microspores is heterogeneous (12). The exine wall with a porous and spinous structure is surrounded with pollenkit and will grow to less than 0.8 µm. Thickness of intine wall varies from 1 to 7.5 µm. This wall consists of two layers that are connected to the exine wall with channels of 0.25 to 1.5 µm diameter (42). The internal layer of intine is important because it will be thickened towards the cytoplasm. During mitosis, microspores produce a large vegetative cell and a spindle shaped reproductive cell. This bicellular stage continues until dispersal of pollens and formation of pollen tube from sperm cells. Pollen cells are able to hydrolyze starches and reserve lipids in their cytoplasm. However, some pollens may use starch as

Fig. 7-2 Longitudal section of saffron ovule with a group of megaspores (x 920).

their cytoplasmic reserves. During dehiscence of anther, size of spherical or ellipsoid mature anthers varies from 45-100 μm. Mature pollens have cytological differences. Pollens with lipid reserves have more intensive cytoplasm (62%) compared to poor cytoplasm of starchy pollens (38%). Some of these starchy pollens contain callus tissues that show their unorganized cytoplasm.

7-4 POLLEN VIABILITY

Based on cytochemical essays, much of pollen has biological activity when released from the anther. However, only a limited number of pollen can successfully germinate. Under *in vitro* conditions, the maximum pollen germination of 20% was obtained in culture media containing sucrose and boron (23, 25). However, mean *in vivo* germination ability of pollens is about 50% and remains constant for a few days after their dispersal.

Saffron pollen encounter dehydration immediately after dehiscence. This process is too fast during the first hours but in temperatures of 18-20°C and relative humidity of 55-70%, water content of pollen reaches an equilibrium within 8-10 hours

(Figure 7-3a). Intensity of dehydration depends on morphological and cytological characteristics of pollen and therefore, varies between different species of *Crocus* (19). Dehydration pattern is strongly related to the porosity of exine wall of pollen that changes from 30-45%. Developmental disorders caused by triploidy of saffron and immature structure of sporoderm or protoplast facilitates dehydration process leading to loss of pollen germination ability after 3 weeks (Figure 7-3b).

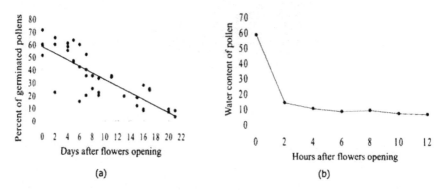

Fig. 7-3 (a) Dehydration of pollen (presented as % water content of pollen) in temperatures of 18-20 °C and relative humidity of 55-75%. (b) Decrease of pollen germination ability over time.

Despite its genetic imbalance, saffron pollen has a long life-span with good germination ability. In fact its cytoplasmic wall is well adapted to fast dehydration after pollen shedding (19). However, the growth of pollen tube usually fails (25) which is comparable with senescence of pollen in the fertile species of *Crocus* (20).

7-5 MEGASPOROGENESIS AND DEVELOPMENT OF EMBRYO SAC

Within the ovule primordium, megasporocytes are surrounded by loculi and nucellar tissues. During meiosis megasporocytes will transform to tetrads or polyade megaspores (Figure 7-1) (11). First meiosis is usually transversal but the second division could be transversal or diagonal. Therefore, megaspores are different both in their size and form (34). Diagonal divisions usually occur in polyades. During the first meiosis the last remaining megasporocyte with chalazal polarization will develop but at the same time micropolar megaspores will disappear. Embryo sac is formed from its penultimate chalazal megaspore but the last megaspore may survive close to embryo sac or develop to form two adjacent sacs (11, 27). In 60% of ovules, functional megaspores follow three nuclear divisions to produce an embryo sac with seven cells including micropolar zygote cells, a central cell and three antipodals. In the remaining 40% of ovules embryo sac is not functional and its development ceased at the first stage.

7-6 OUTCROSSING

Saffron pistil consists of a) an ovary with three carpels that contain approximately 29 inverse ovules, b) style with 8-9 cm length, and c) style branches that form stigma. Style is unicellular and dry type but without the protein pellicle, with pubescent margins (42)

Following pollination, saffron pollens adhere to the trichomes and germinate after rapid hydration. During 50-70 minutes after pollination pollen tubes pass through the stigma and grow within the style channel. Saffron style has three channels separated with long cells that provide polysacharides for pollen tubes. Growth of pollen tubes within the style is usually ceased and only few tubes will be able to enter the ovary (25). The fertilization route is located across the axile and connected to the ovule micropyle. This route is surrounded with pear shaped cells (Figure 7-5) and produces gylcoproteins, is of great importance for growth of pollen tube in *Crocus* species (17, 22). The behavior of these pollen tubes is similar to self-fertilizing species of *Crocus* and their growth will be stopped within axiles (Figure 7-4). In any case, the fertilized ovules will never produce seeds.

7-7 INTERSPECIFIC CROSSING

Saffron pistil supports germination and growth of pollen tubes even after inter pollination. Growth of pollen tubes from crossing between

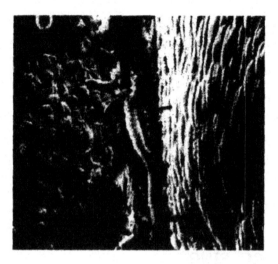

Fig. 7-4 Electron microscope photo of axiles below the cuticule layer of saffron ovary (C). Arrows point to pear shaped cells and pollen tube (arrow = 100 μm).

Fig. 7-5 Longitudal section of saffron ovule 25 days after crossing with *Crocus shadriatiaus*. Arrow shows three developed embryo cells, OI = external and II = internal epidermis (x 140).

different species of *Crocus* is quite similar to self-pollinated species. However, if the segregated descents were not *Crocus sativus* growth of pollen tube will cease before reaching ovule (61). In crosses between *Crocus sativus* and species such as *C. thomassi* Ten; *C. hadriaticus* Herb.; and *C. oreoceticus* B.L. Burrt., the pollen tube reaches the ovary and some fertilization occurs (Figure 7-5) (20). The highest percentage of *C. sativus* ovaries containing zygote and endosperm were obtained after crossing with *C. thomassi* (16.8%), *C. oreoceticus* (11.8%) and *C. hadriaticus* (8.5%). Embryological studies on *C. sativus* have shown that only a few numbers of fertilized ovaries are able to produce mature seeds (15) and abortion is often during embryo development. From 125 crossed flowers of *C. thomassi* only 47 mature seeds were obtained. Half of them were germinated and developed into a mature plant with fruits, seeds and embryos larger than *C. sativus*.

Inter-specific crossing with *C. sativus* as pollen donor, is usually unsuccessful (25).

7-8 *IN VITRO* PROPAGATION AND PRODUCTION OF SECONDARY METABOLITES

Saffron is a geophyte herb with triploid nature that is widely used as an aromatic medicinal plant with valuable secondary metabolites. Saffron growth starts from a corm and a thickened stem that provides permanent reserves for germinating buds. New corms are produced from swollen buds of maternal corm. By the end of season these corms are surrounded with thin dried leaves (79).

Two types of leaves are produced from growing corm buds, true and cataphylus leaves that protect young true leaves. The total number of leaves per bud is about 12-14. Intensive adventitious roots are produced at the base of growing buds and help better establishment of plant.

Saffron is a sterile triploid plant and therefore, corms are used for its vegetative propagation. Bacterial, fungal and viral diseases usually infect corms and remain active after harvesting. Despite care in sanitation, these pathogens are the main cause of necrosis of corm and young leaves and consequent decrease in flowering. Plants infected by fungal or bacterial pathogens could be treated with appropriate chemicals but such treatments are not effective in viral infections (60).

Meristem tip culture and plant regeneration from the cultured tissues is the only way for wide production of pathogen free saffron (28, 50). Tissue culture techniques have been used for many geophytes including saffron. The technique is based on totipotency or the ability of plant intact cells, tissues and organs to develop new organs or somatic embryos when grown in a specific culture medium (55).

Explants from corm tissues, lateral or apical buds, leaves and nodal tissues, and different floral parts have been used for *in vitro* regeneration of saffron (see Table 7-1 for details). The results depend on plant age, selected organ, and composition of medium and particularly amount and composition of growth regulators.

Saffron is a sterile plant and classical methods cannot be used for its breeding. However, *in vitro* culture of protoplast, anther as well as gene transfer techniques could be used for genetic breeding of saffron.

7-8-1 Composition of Culture Media

Saffron explants have been grown and regenerated in solid agar medium and solution cultures. The most common media are MS, LS, N_6, W, and B_5 (Table 7-1). In some research better growth was reported by reducing minerals to their half concentration.

Concentration of growth regulators depends on the plant organ and growth stage. It has been reported that culture initiation (stage 1) or proliferation (stage 2) require quite different media compared to hardening or root and corm formation (stage 3).

Growth regulators that are used for *in vitro* saffron culture were auxins including NAA, IBA, IAA, 2,4-D, cytokinines such as KN, BAP, ZN as well as GA3, ABA, and ethylene at different concentrations. Sucrose content varies from 2-12% and other compounds such as charcoal, coconut milk and ascorbic acid were also reported.

Table 7-1   *Summary of in-vitro studies on saffron (Crocus sativus L.)*

Source of explants	Growth medium a	Growth regulators (Micromolar)	Additives (g/liter)	Stage[b]	Morphogenesis response	References
Lateral buds	Macro Ms, Micro, Vitamin N6	(6/9) ZN, (10/3) BAP (0/1) GA3, (0/05) NAA	Sucrose (40)	First	Bud 's growth	3
Lateral buds	MS (1/2 N concentrate)	(6/6) BAP	Sucrose (30)	Second	Branch propagation	3
Branches (*in vitro*)	MS (1/2 N concentrate)	-	Sucrose (60)	Third	Corm, Branch	3
Shoot meristem	MS	(20) NAA, (20) BAP	-	First	Somatic embryogenesis	4
Somatic embryos	1/2 MS concentrate	(57/8) GA3	-	Second	Mature embryos	4
Mature embryos	1/2 MS concentrate	(5) NAA, (5) BAP	Active coal (20)	Third	Petal, Corm	4
Tip of flower buds on young corms	MS	(9/3) KN, (9) 2,4-D	-	First	Callus	30
Callus	MS	(9/3) KN, (5/8) NAA	-	Second	Corm, Branch, floral structures	-
Small corm parts	MS	(2/3) KN	-	First	Bud	31
Small corm parts	MS	(4/5) 2,4-D	-	Second	Small corm	31
Corms	MS	(5/7) IAA, (5/3) NAA	-	First, Second, Third	Callus, seedling	32
Other parts of flower	W	(10/7) NAA, 11/4 (ZN)	Sucrose (30), Coconut milk (20), Glutamine (20)	First, Second	Stigma structure	35

(Table 7.1 Contd.)

(Table 7.1 Contd.)

Source of explants	Growth medium a	Growth regulators (Micromolar)	Additives (g/liter)	Stage	Morphogenesis response	References
Petals	W	(21/5-43) NAA	Sucrose (30), Coconut milk (20), Glutamine (20)	First	Callus	37
	W	(2/2-36/2) 2,4-D, (2/2-35/5)	Sucrose (30), Coconut milk (20),	First	Callus	37
Anthers	W	(22/8) ZN, (43) NAA	Sucrose (30), Coconut milk (20), Glutamine (20)	First	Callus	37
Disrupted ovary	W	(5/6-22/8) ZN, (10/7-43) NAA	-	First	Stigma structure	38
Callus	MS	(2/3) KN, (9) 2,4-D	Ascorbic acid (0/1)	Second	Somatic embryogenesis	37
Spherical embryos	MS (1/2 N concentrate)	(9/3) KN, (10/7) IAA	-	Second, Third	Branch, seedling	37
Meristem bud	MS	(18/2) ZN, (21/4) NAA	-	First	Callus	39
Spherical callus	MS	(3/8) ABA	-	Second	Branch	39
Corm and bud slices	MS			Third	Corm production	44
Corm and bud slices	MS	(9/3) KN, (11/4) IAA	Ascorbic acid (0/1)	Third	seedling	44

(Table 7.1 Contd.)

(Table 7.1 Contd.)

Other parts of flower	MS or N6	(22/2-26/6) BAP, (11/4) IAA	-	First	Callus	45
Flower tissue slices	MS or N6	(22/2-26/6) BAP, (26/8) NAA	-		Stigma & Style structure	46
Stigma structures				First	Crocin, picrocrocin and safranal synthesis	47
Corm parts	Different	(4/6) KN, (53/7) NAA	Sucrose (20)	First	No growth	48
Corm parts	Minerals B5+ organic materials MS	(0/4) KN, (4/5) 2,4-D	-	Second	Micro corm	-
Base of leaves Callus	N6 MS	(9) 2,4-D (2/2) BAP, (9) 2,4-D	Sucrose (30)	First First, Second	Callus Bud	49 49
Buds	MS (1/2 N concentrate)	(2/2) BAP, (1/1) NAA	-	First	Branch	49
Corm	MS	(5/7) IAA	Coconut milk (20)	First	Callus	52
Replanted branches	MS	(2/2) BAP, (2/3-4/5) 2,4-D	Coconut milk (20)	Third	Plantlet on original explants	52
Corm	MS	(2/2) BAP, (2/3-4/5) 2,4-D	Coconut milk (20)	First	Callus	53
Callus	Liquid MS	(2/2) BAP, ZN, 2,4-D	Coconut milk (20)	Second	enlargement	53
Node	MS	(4/5) 2,4-D	Coconut milk (20)	Second	enlargement	53
Spherical Nudes	MS	(4/4) BAP, (5/5) NAA	Coconut milk (20)	Third	Branch	53

(Table 7.1 Contd.)

(Table 7.1 Contd.)

Source of explants	Growth medium	Growth regulators (Micromolar)	Additives (g/liter)	Stage[b]	Morphogenesis response	References
Callus culture, Harvested Stigmata	LS	(13/3) BAP, (0.5) NAA	Coconut milk (20)		Stigma structure	56
Shoot, young petals	MS	(50) NAA, (30) BAP	Coconut milk (20)		Stigma & Style structure	59
Stigma & Style structure	MS	(21/4) NNA, (3/2) KN	Coconut milk (20)	First, Second	Stigma structure	59
Other parts of corm, bud	MS	Different	Different	First	Callus from corm parts	62
Apical and lateral buds and their callus explants	MS	(4/6-13/6)ZN, (2/6-16/1)NAA or IAA	Sucrose (30)	Second	Corm	63
Other parts of bud, flower, petal and ovary	B5, LS		Sucrose (50-120)	First, Second	Stigma structure	65
Flower, petal or ovary	MS	(8/9-22/2) BAP, (0/5-5/4) NAA			Stigma structure	68
Disrupted apical buds or buds on small corms	MS	(4/5) 2,4-D, (13/7) ZN				
Pre treatment of explants with ethylene	-			First, Second Third, Second	Leaf and corm production	72

(Table 7.1 Contd.)

(Table 7.1 Contd.)

Explant	Medium	Growth regulators	Additives	Development stages	Response	Ref.
Young stigmata, ovaries or those combination	LS	(23/2, 4/6) KN (3/2, 0/3) BAP		First	Stigma growth, Crocin biosynthesis	74
Meristem tissue	N6, LS	(54,0/5) NAA, (23/2-4/6) KN, (49,0/6) IBA		First, Second	Stigma structure	74
Stigma	MS	(3/2)BAP, (54)NAA		First, Second	Stigma structure, Crocin, picrocrocin	75
Ovary	MS	(54 or 27) NAA	-	First, Second	Stigma structure, Crocin, picrocrocin	76
Corm	MS	(2/3) KN, (9) 2,4-D	Sucrose (30)	First, Second	Callus	77
Flowers buds	MS (cell culture)	(2/3) KN, (9) 2,4-D	Sucrose (30)	First, Second	Spherical red globule shape cultures, crocin, crocetin, picrocrocin	78
Halved immature ovary	B5+MS (Organic matter)	(44/4) BAP, (5/4) NAA	Hydrolyzed casein (0/05%) and L-Alanine (11/2 milimolar)	First, Second, Third	Stigma structure, root, corm, petal, leaf	58
Corm	MS	(9/3) KN, (9) 2,4-D (5/4) NAA, (9/3) KN	-	First First, Second, Third	Corm, branch, flower	32

[a] N= Nitrogen; W= White (1963); N6= Nitsch and Nitsch (1969); LS= Linsmaier and Skoog (1965); B5= Gambourg (1968); MS= Murashige and Skoog (1962)

[b] Development stages: First = initial; Second = regeneration or propagation; Third = hardening, root and corm formation.

7-9 BUD AND MERISTEM CULTURE

Buds and meristem, tissues of saffron corm have been used as explants for *in vitro* propagation. During development and differentiation, explants follow organogenesis or somatic embryogenesis stages. Shoot meristems with length of 8-10 mm together with two leaf primordiums isolated from small corms produced callus in LS medium containing BAP and NAA. The callus were developed to spherical cells that differentiated to somatic embryos when grown in half concentration MS medium without growth regulators. Plant regeneration from embryos was only possible in presence of BAP, NAA and charcoal (4). In a similar experiment, callus was produced from explants of corm meristems grown in a medium containing 2, 4-D and differentiated to spherical embryos when NNA and KN or ascorbic acid were added. However, further growth of embryos was possible in MS solution culture with ABA and minerals at half concentration. These embryos formed adventitious buds that regenerated to mature plants in solution culture (39). It has been reported that both lateral and apical buds are proper explants for *in vitro* shoot regeneration. Lateral buds of mature corms produced 6-8 young shoots in half concentration MS medium with BAP. Micro corms were formed from these shoots in the absence of growth regulators under short days and temperature of 15°C (4).

Dormant lateral or apical together with parts of corm tissues were differentiated to shoot and micro corm in presence of IAA and ZN when grown at 10°C and long days condition. Regeneration rate was 3 micro corms after two months, that is, relatively low for *in vitro* propagation. Callus of the same explants were differentiated to embryos when growth regulators were added. However, the embryos had a bipolar structure similar to dicotyledons with appearance like shoot and root tip meristem (62).

7-10 CORM AND LEAF EXPLANTS

Storage tissues of saffron corm have been reported by researchers as potential explants but the detailed characteristics of such an explant was not described. Nodal tissues with their high division activity and regeneration ability could also be used as explants.

Ding et al. (31) successfully obtained callus from corm tissues and regenerated seedlings in the presence of NAA and IAA. When corm parts were used, micro corms were formed on explants grown in medium containing 2, 4-D (48). Formation of callus and development of bud and seedling was reported when coconut milk together with 2, 4-D and BAP were added to growth medium of corm explants (52). The same result was obtained from corm explants in presence of 2, 4-D and ZN. Substitution of ZN with BAP led to differentiation of shoots from callus after three months (53). Plessner et al. (72) used small corms for developing shoot from their apical and lateral buds. When the small corms were treated with ethylene and their apical buds scratched, micro corms were formed in a medium containing 2, 4-D, KN and ZN (see Table 7-2 and Figure 7-6 for details). There is only one report where callus was formed from leaf explants in presence of NAA and BAP. After 8 months several buds were formed that developed into shoots when minerals of MS medium were reduced to half concentration and IAA was added (49).

Table 7-2 *Effects of ethylene, ethphone and scratching apical buds of in vitro grown small corms on development of lateral buds after 12 weeks*

Treatment	Formation and development of buds	Number of leaves
Control[a]	1± 0	5
Micro scratching[b]	4 corms	-
Ethphone (1000 mg lit^{-1} for one hour)[c]	7± 3 d	-
Ethylene (1000 mg lit^{-1} for one hour)[d]	6 ±2 d	-
Ethylene (1000 mg lit^{-1}) + scratching	15± 3 d	-

[a]1 ml lit^{-1} 2,4-D was added.
[b]Scratching with surgical razor.
[c]Ethylene or ethephone were used as pre-treatment.
[d]Corm or micro corm was formed on the base of apical bud.

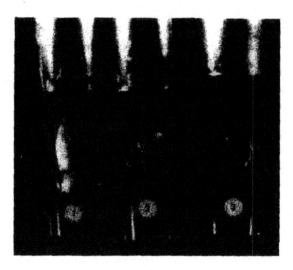

Fig. 7-6 Development of buds on *in vitro* grown corms of saffron after pertreatment with 1) ethylene and ethphone; 2) water (control); 3) 1000 ppm ethphone; 4) 1000-ppm ethylene (see Table 7-2).

7-11 EXPLANTS FROM FLORAL PARTS

Morphogenic response of different flower parts such as ovary, style, stamen or stigma depends on flower age and development stage of the separated part. In many cases it has been reported that the callus formed in presence of growth regulators were differentiated to structures similar to style or stigma with yellow-orange pigments.

Halved ovaries in medium containing NAA and ZN (9) and intact ovary and stigma in presence of NAA and BAP (37, 38) were differentiated to stigma or corm. Formation of up to 75 stigmas with yellow-orange color was reported in medium with NAA and IBA (35). Similarly, stigma was differentiated from intact flowers (13).

Formation of pigmented structures was enhanced when sucrose (5-10%) together with BAP, NAA and alanine were added (68). In LS medium with NAA and BAP, these structures grew to 15 mm after 3-4 months (23).

Himeno et al (46) reported that stigmas were differentiated from saffron carpel tissues grown in LS medium containing NAA and KN with 10:1 ratio. The regenerated stigmas were biologically mature and were able to receive pollen.

Recently, explants from halved immature ovaries were used in different base medias with different proportion of growth regulators to evaluate frequency of formation of stigmas (58). Results showed that after 50 to 60 days structures similar to stigma were successfully formed in B5 medium. The medium contained NAA (5.43 μm), BA (44 μm), organic compounds of MS medium, hydrolyzed cazeine (5%) and L-alanine (11.2 μm).

7-12 *IN VITRO* PRODUCTION OF SECONDARY METABOLITES

Explants from floral parts, organs or their callus are the main sources of production of metabolites from saffron. These metabolites are products of flower buds or re-differentiation of style-like or stigma-like structures.

Re-differentiated organs are usually pigmented with colors varying from yellow to orange and red. Extracts of these organs contain crocin, picrocrocin, crocetin and in some cases safranal. Amount and quality of saffron metabolites depends on type of *in vitro* differentiated tissues or organs. In stigma-like structures that were formed *in vitro* culture of floral parts the amount of crocin and picrocrocin were respectively, 6 and 11 times lower than natural stigma. Sensory analysis has shown that *in vitro* produced metabolites were different from products of natural flowers (75, 76).

When stigma and ovaries were grown in a specific medium, three yellow pigments were produced from differentiated stigma-like structures but their amount was lower than the same pigments in natural conditions (74). Stigmas that were developed from halved ovaries produced an orange pigment similar to natural saffron metabolites (35).

In red and filamentous structures differentiated from flower buds the amount of crocin and picrocrocin were higher that normal stigmas and their safranal content was comparable with that of natural saffron (78). The amount of crocin, crocetin glucosil esters, picrocrocin and safranal in style-like and stigma-like structures differentiated from culture of halved immature ovaries were measures using HPLC. Results showed that the amount of these metabolites was quite similar to natural metabolites of saffron (26).

In recent years *in vitro* synthesis of some saffron metabolites and their biochemical conversion have been studied. As an example, conjugated sugars with anti-cancer effects has been isolated from callus culture of saffron corms (33). Biochemical pathways for metabolites such as crocetin monoglucosyl and diglucobil in saffron tissue cultures have been studied using electrophoresis method. These results are the basis for production of synthetic compounds similar to saffron metabolites (36, 37).

7-13 CONCLUSION AND FUTURE PROSPECTS

In *Crocus sativus* due to its triploid genome, transition from sporophytic to gametophytic stage is characterized by cytological aberrations. Therefore the produced spores are genetically and cytologically abnormal. Pollens are identical in size, shape and in development of entine wall and cytoplasm. In embryo sac up to 40% of ovules are defective. However, the sterility of saffron cannot be fully described by these disorders. It seems that incompatibility mechanisms that prevent self-breeding in fertile species of *Crocus* (16, 17) are involved in saffron ovary. Integration of saffron genome with genetic traits of its wild relative species could lead to stability of traits such as resistance to fungal and viral diseases (72), induction of hysternathy (72) and improvement of yield and quality of saffron metabolites (66). It seems that C. *thamasii* is the best species for hybridization programs (18, 70). This species is abundant in south Italy and west Yugoslavia. The shortcoming in saffron hybridization programs is the need for high number of hybrid seeds for selection experiments and difficulties in normal development of seeds. However, experimental results (15, 24) have shown that *in vitro* culture of cross-pollinated ovaries of Crocus could be a successful way for seed production.

Genetic improvement using DNA manipulation and gene transfer are new biotechnological approaches to overcome saffron sterility.

7-14 SUMMARY

Saffron has a triploid genome that causes meiotic abnormalities and consequently variation in sporogenesis and gametogenesis. Therefore, development potential of megasporocytes is strongly limited. However, reproduction system of saffron, like fertile species of *Crocus* provides the possibility of introgression with relative species. This ability together with using *in vitro* techniques has led to successful seed production as a new approach for genetic improvement of saffron. Saffron has a low vegetative propagation rate. However, using tissue culture techniques and regeneration of somatic embryos, fast propagation and production of virus free corms from differentiated embryos would be possible. Micro corms have been regenerated from shoot apical and lateral buds in medium containing sucrose and in absence of growth regulators. Corm tissues could also be used for regeneration of seedlings or micro corms. Yellow-orange structures similar to stigma have been formed in a medium containing growth regulators. Crocin and picrocrocin content of these structures were lower or the same as natural stigmas.

References

1. Aghamohammadi, Z. 1976. Evaluation of sexual reproduction, diversification by mutagens and sterility of saffron (*Crocus sativus* L.) by measuring direct pollen growth and chromosomes coupling. MSc thesis, Tehran University, Iran.

2. Aghayev, Y. 1993. Some basic problems in genetics, cytogenetics and selection of saffron (*Crocus sativus* L.). Abstracts Book of The Second

National Symposium of Saffron and Medicinal Plants, Gonabad, Iran. P: 12-13.

3. Aguero, C. and R. Tizio, 1994. *In vitro* mass bulbiftcaiion as a preliminary contribution to saffron (*Crocus sativus* L.). Biocell, 18: 55.63.

4. Ahuja, A., S. Koul, and G. Ram, 1994. Somatic embryogenesis and regeneration of plantelts in saffron, *Crocus sativus* L. Indian Journal of Experimental Biology, 32: 135-140.

5. Bagheri, A., M. Ghafari and M.H. Rashed Mohassl, 1988. Evaluation of saffron diversity and its use in breeding programs. Scientific and Industrial Research Organization of Iran, Khoran Center.

6. Brandizzi, F. and M.G. Caiola. 1998. Flow cytometric analysis of nuclear DNA in *Crocus sativus* and allies (Iridaceae). Plant Syatematics and Evolution, 211(3-4): 154-164.

7. Brighton, C.A. 1977. Cytology of *Crocus sativus* and its allies (Iridaceae). Plant Systematics and Evoluton, 128: 137-157.

8. Carroll, C.P. 1966. Autopolipoidy and the assortment of chromosomes. Chromosoma (Berl.), 18: 19-43.

9. Chichiricco, G., and M. Grilli Caiola, 1986. *Crocus sativus* pollen germination and pollen tube growth *in vitro* and after intraspesitic and interspecific pollination. Canadian Journal of Botany, 64: 2774-2777.

10. Chichiricco, G. 1984. Karyotype and meiotic behavior of the triploid *Crocus sativus* L. Caryologia, 37: 233-239.

11. Chichiricco, G. 1987. Megasporogenesis and development of embryo sac in *Crocus sativus* L. Caryologia, 40: 59-69.

12. Chichiricco, G. 1989. Microsporogenesis and development in *Crocus sativus* L. Caryologia, 42: 237-249.

13. Chichiricco, G. 1989. Embryology of *Crocus thomasii* (Iridaceae). Plant Systematics and Evolution 168: 39-47.

14. Chichiricco, G. 1989. Fertilization of *Crocus sativus* ovules and development of seeds after stigmatic pollination with *Crocus thomasii* (Iridaceae). Giornale Botanica Italiano, 123: 31-37.

15. Chichiricco, G. 1990. Fruit and seed development of cultured fertilized ovaries of *Crocus*. Annals of Botany, 48: 87-91.

16. Chichiricco, G. 1993. Pregamic and postgamic self-incompatibility systems in *Crocus* (Iridaceae). Plant Systematics and Evolution, 185: 219-227.

17. Chichiricco, G. 1996. Intra and interspecific reproductive barriers in Crocus (Iridaceae). Plant Systematics and Evolution, 201: 83-92.
18. Chichiricco, G. 1999. Development stages of pollen wall and tapetum in some Crocus species. Grana, 38: 31-41.
19. Chichiricco, G. 2000. Dehydration and viability of saffron (*Crocus sativus*). Grana, 39: 275-278.
20. Chichiricco, G. 2000. Viability-germinability of Crocus (Iridaccae) pollen in relation to cyto-and ecophysiologycal factors. Flora, 195: 193-199.
21. Chichiricco. G. 2000. Ripening and dehiscence of the anther in some Crocus (Iriaceae) species. Caryologia, 53: 255-260.
22. Chichiricco, G. P. Aimola, and A.M. Ragnelli, 1995. Cytochemical and ultrastructural study of the ovarian transmitting tract of Crocus (Iridaceae). Giornale Botanico Italiano, 129: 21.
23. Chichiricco, G. and M. Grilli Caiola. 1982. Germination and viability of the pollen of Crocus sativus L. Giornale Botanico Italiano, 116: 167-173.
24. Chichiricco, G., and M. Grilli Caiola. 1987. *In vitro* development of parthenocarpic fruits of Crocus sativus L. Plant Cell Tissue and Organ Culture, 11: 75-78.
25. Chichiricco, G. and M. Grilli Caiola, 1984. Crocus sativus pollen tube growth in intra- and interspecific pollination. Caryilogia, 37: 115-125.
26. Cute, F., F. Comier, C. Dufresne, and C. Willemot. 2001. A highly specific glucosyltranferase is involved in the synthesis of crocetin glucosylesters in Crocus sativus cultured cell. Journal of Plant Physiology, 158: 553-560
27. De Nettancourt, D. 1977. Incompatibility in Angioperms. Springer, Berlin, Heidelberg, New York.
28. Debergh, P.C. and P.E. Read. 1991. Micropropagation. In: Debergh, P.C. and R..H. Zimmerman, (eds.) Micropropagation: Technology and Application. Kluwer Acad. Pub. Dordrecht, pp, 1-13.
29. Dhar, A.K., R. Sapru and K. Rekha. 1988. Studies on saffron in Kashmir, a. Variation in natural population and its cytological behaviour. Crop Improvement, 15: 48-52.
30. Dhar, A.K. and R. Sapru. 1993. Studies on saffron in Kashmir. III. *In vitro* production of corm and shoot - like structures. Indian Journal of Genetics and Plant Breeding, 53:193-196.

31. Ding, B.Z., S.H. Bai, Y. Wu and X.P. Fan. 1979. Preliminary report on tissue culture for corm production of *Crocus sativus* L. Acta Botanica Sinica, 21: 32-41.

32. Ding, B.Z. 1981. Induction of callus and regeneration of plantlets from the corm of *Crocus sativus* callus extract. Acta Botanica Sinica, 23: 419-420.

33. Dufresne, C., P. Cormier, and S. Dorion. 1997- In *vimo* formation of crocetin glucosyl esters by *Crocus sativus* callus extract. Planta Medica, 63: 150-153.

34. Dufresne, C., F. Cormier, S. Dorion, U.A. Niggli, S. Pfister and H. Pfander. 1999. Glycosylation of encapsulated crocetin by a *Crocus sativus* L. cell culture. Enzyme and Microbial Technology, 24: 453-462.

35. Escribano, J., A. Piqeras, J., Medine, A. Rubio, M. Alvarez - Oorti and J.A. Frenandez. 1999. Production of cytotoxic proteoglycan using callus culture of saffron (*Crocus sativus* L.). Journal of Biotechnology, 73:53-59.

36. Estilaee, A. 1978. Difficulties of seed production in saffron and production of new cultivars using mutation. Pajohandeh, 21: 242-272.

37. Fakhari, F. and P.K.. Evans. 1990. Morphogenic potential floral explants of *Crocus sativus* L. for *in vitro* production of saffron. Journal of Experimental Botany, 41:47-52

38. Gambourg, O.L., R.A. Miller and G, Ojima. 1968. Nutrient requirements of suspension culture of soybean root cells. Experimental Cell Research, 50:148-151.

39. George, P.S., S. Visvanath, G.A. Ravishankar and L.V. Venkataraman. 1992, Tissue culture of saffron (*Crocus sativus* L.): Somatic embryogenesis and shoot regeneration. Food Biotechnology, 6: 217-223.

40. Ghafari, S.M. 1980. Cytogenetic studies of cultivated saffron and its breeding. Center of Biochemical and Biophysical Research, Tehran University, Iran.

41. Ghaffari, S.M. 1986. Cytogenetic studies of cultivated *Crocus sativus* (Iridaceae). Plant Systematics and Evolution, 153: 199-204.

42. Grilli Caiola, M., M. Castagnola and G. Chichiricco, 1985. Ultrastructural study of saffron pollen. Giornale Botanico Italiano, 119: 61-66.

43. Grilli Caiola, M., and G. Chichiricco. 1990. Stuctural organization of the pistil in saffron (*Crocus sativus* L.). Israel Journal of Botany, 40: 199-207.

44. Gui, Y.L., T.Y. Xu, S.R. Gu, S.Q. Liu, G.D. Sun and O. Zhang, 1988. Corm formation of saffron Crocus *in vitro*. Acta Botanica Sinica, 30: 338-340.

45. Han L.L. and X.Y. Zhang, 1993. Morphogenesis of style stigma like structures form floral explants of Crocus sativus L. and identification of the pigments. Acta Botanica Sinica, 35: 157-160.

46. Himeno, H., H. Matsushima and K. Sano. 1988. Scanning electron microscopic study on the *in vitro* organogenesis of saffron stigma and style - like structures. Plant Science 58: 93-101.

47. Himeno, H. and K. Sano. 1987. Synthesis of crocin, picrocrocin, and safranal by saffron stigma - like structure proliferated *in vitro*. Agricultural and Biological Chemistry, 51: 234-239.

48. Homes, J., M. Legros and M. Jaziri. 1987. *In vitro* multiplication of Crocus sativus L. Acta Horticulturae, 212: 675-676.

49. Huang, S.Y. 1987. A study on tissue culture of Crocus sativus. Plant Physiology Communication, 6: 17-19.

50. Hussey, G. 1975. Totipotency in tissue explants and callus of some members of the Liliaceae, Iridaceae and Amaryllidaceae. Journal of Experimental Botany, 26: 253-262.

51. Hussey. G. 1986. Problems and prospects in the *in vitro* propagation of herbaceous plants. In: Whithcrs, L-A, and P.G. Alderson (eds). Plant Tissue Culture and its Agricultural Application. Butterworths, London, pp. 69-.84.

52. Ilahi, Jabbeen, M. and N. Firdous. 1987. Morphogenesis from the callus of saffron (Crocus sativus). Japanese Journal of Breeding, 38: 371-374.

53. Isa, T. and T. Ogasawara. 1988. Efficient regeneration from the callus of saffron (Crocus sativus). Japanese Journal of Breeding, 39: 261-272.

54. Karasawa, K. 1933. On the tripoloidy of Crocus sativus L. and its high sterility. Japanese Journal of Genetics, 9: 6-8.

55. Kim, K.W. and A.A. De Hertogh. 1996, Tissue culture of ornamental flowering geophytes. Horticultural Review, 18: 87-169.

56. Koyama, A., Y. Ohmori, N. liujioka, H. Miyagawa, K. Yamasaki and H. Kohda. 1988. Formation of stigma-like structures and pigment in cultured tissues of Crocus sativus. Planta Medica, 54: 375-376.

57. Linsmaier, E, and F. Skoog, 1965. Organic growth factors requirements of tobacco tissue cultures. Physiologia. Plantarum, 18: 100-127.

58. Loskutov, A.V., C.W. Beninger, T.M. Ball, G.L. Hosfield, M.Nair and K.C. Sink.1999. Optimization of *in vitro* conditions for stigma-like structure production form half - ovary explants of *Crocus sativus* L. *In vitro* Cellular and Development Biology – Plant,. 35: 200-205.

59. Lu, W.L, X.R, Tong, Q. Zhang and W.W. Gao. 1992. Study on *in vitro* regeneration of stigma like structure in *Crocus sativus* L. Acta Botanica Sinica, 34: 251-252.

60. Magie, R.O. and S.L. Poe. 1982. Disease and pest associates of bulbs and plants. In: Koeing, N. and. W. Crowley (eds). The World of Gladiolus (N.) North American Gladiolus Council, Edgerton Press, MD, pp. 155-156.

61. Mathew, B. 1977. *Crocus sativus* and its allies (Iridacae). Plant Systematic and Evolution, 128: 89-103.

62. Milyaeva, E.L,, E.V. Komarova, N.S. Azizbekova, D. D. Akhundovam and R. G. Butenko (eds). 1988. Features of morphogenesis in *Crocus sativus in vitro* culture. Biol. Kultiviruemykh Kletok. Bioleklmclogiya, 1: 146.

63. Milyaeva, E.L., N.S. Azizbekova, E.N, Komarova and D.D. Akundova. 1995. *In vitro* formation of regenerant corms of saffron crocus (*Crocus sativus* L.). Russian Journal of Plant Physiology, 42:112-119.

64. Mutashige, T. and M.F. Skoog. 1962. A revised medium of rapid growth and bio-essay with tobacco tissue cultures. Physiologia Plantarum, 15: 473-497.

65. Namera, A., N. Koyoma, K.A. Fuji, H. Yamasaki and H.Konda. 1987. Formation of stigma like structures and pigments in cultured tissues of *Crocus sativus*. Japanese Journal of Pharmcognosy, 41: 260-262.

66. Negbi, M., B. Dagan, A. Dror and D. Basker. 1989. Growth flowering, vegetative reproduction and dormancy in the saffron crocus (*Crocus sativus* L.). Israel Journal of Botany, 38: 95-113.

67. Nitsch, J.P. and C. Nitsch. 1969. Auxin - dependent growth of excised Hellianthus tissues. American Journal of Botany, 43: 839-851.

68. Otsuka, M., H.S. Saimoto. Y. Mutata and M. Kawashima. 1992. Methods for producing saffron stigma - like tissue. United States Patent. US 5085995,8 pp, A28.08.89 US 399037, P 04.02.92.

69. Pacini, E. 1990. Tapetum and microspore function. In: Blackmore, S. and R.B. Knox, (Eds). Microspores, Evolution and Ontongeny. Academic Press, London, pp. 213-237.

70. Paradics, M. 1957. Osservazioni sulla costiuzione eciclo di sviluppo di *Crocus sativus thomasii* Ten. Nuovo Giornalc Botanico Italiano, 64: 347-367.

71. Plessner, O., M. Negbi, M. Ziv and D. Basker. 1989. Effects of temperature on the flowering of the saffron crocus (*Crocus sativus* L.): Induction of hysteranthy. Israel Journal of Botany, 38: 1-7.

72. Plessner, O., M. Zi. and M. Negbi. 1990. *In vitro* corm production in the saffron (*Crocus sativus* L.). Plant Cell Tissue and Organ Culture, 20: 89.94.

73. Russo, M., C.P. Martelli, M. Cresti and F. Ciampolini. 1979. Bean yellow mosaic virus in saffron. Phytopathological Mediteranea, 18: 189-201.

74. Sano. K. and H.Himeno. 1987. *In vitro* proliferation of saffron (*Crocus sativus* L.). Plant Cell, Tissue and Organ Culture, 11: 159-166.

75. Sarma, K-S-, K- Maesato, H.T. Hala and Y. Sonoda, 1990. *In vitro* production of stigma-like structures from stigma explants of *Crocus sativus* L. Journal of Experimental Botany, 41: 645-648.

76. Sarma, K.S., K.Sharada, K.Maesato, H.T, Hara And Y. Sonoda. 1991. Chemical and sensory analysis of saffron produced through tissue cultures of *Crocus sativus*. Plant Cell Tissue and Organ Culture, 26: 11-16.

77. Sarma, K.S., G.A. Ravishankar, L.V. Venkataraman and H.L. Sreenath. 1992. Chromosome stability of callus cultures of *Crocus sativus*. Journal of Spice and Aromatic Crops, 1: 157-159.

78. Visvanath, S., G.H. Ravishankar and L.V. Venkataraman. 1990. Induction of crocin, crocetin, picrocrocin, and safranal synthesis in callus cultures or saffron: *Crocus sativus L*. Biotechnology and Applied Biochemistry, 12; 336-340.

79. Warburg, L.L. 1957. Crocus. Endeavor 16: 209-216.

80. White, P. R. 1963, The Cultivation of Animal and Plant Cells, Ronald Press, New York. VII. 228 pp.

CHAPTER **8**

Economic Aspects of Saffron

A. Karbasi
Faculty of Agriculture, University of Zabol

8-1 INTRODUCTION

Considering the diversity of climate in different part of the world, some areas have relative advantage for production of specific crops. The special geographical characteristics that govern southern and central parts of Khorasan, Iran, such as desertification, poor water quality, water shortage, soil fertility and traditional methods of agricultural production are some of the limiting factors for growth of agricultural crops in these areas.

Saffron can tolerate water shortage in these areas because of its special peculiarities among different crops. Due to very high levels of efficiency, growing this crop keeps farmers from migrating to larger cities. Importing saffron and competition within international market necessitates agricultural economical research on saffron specifically on marketing, and importing saffron.

In Iran, saffron planting is important from different points of view, such as, higher water use efficacy, providing work for labor, and non-petroleum export (3). Presently, the main income of about 85,000 families in southern and central Khorasan is from saffron (11). From the labor perspective, saffron cultivation needs about 270 labor per capita per day for each hectare. Iran is undoubtedly the greatest producer and exporter of saffron in the world and more than 90% of world saffron is produced in Iran (9). In this section different aspects of saffron marketing and price will be discussed. We will also discuss the saffron export trends and problems during recent years in saffron export and marketing.

8-2 SAFFRON MARKETING

Due to the special nature of crops such as putrification, volume, moisture, the number of producers and consumers, and limitation for crop planting and harvesting, the marketing of agricultural crops differs from industrial products. In late 20th and early 21st century with introduction of new technologies into traditional agriculture and commercialization of agricultural enterprises, services and marketing of agriculture have changed time.

Presently, the dried stigma of saffron is packed and presented to internal and external markets. In Iran, dried saffron is packed in hard boxes and supplied in nylon or fabric bags, with little regard for hygiene. These external factors affect saffron quality (13). Loose, unpacked saffron is bought from Iran by Spain and processed and packaged attractively and sold at twice the price in the European Common Market. Various private and commercial vehicles presently transport saffron, with little regard to the high prices for relatively small volumes of the product. A vehicle suited to the needs of transporting saffron needs to be designed. Saffron is not stored in specially designed storages spaces. Usually farmers keep it in their homes without providing special rooms with low humidity and cool temperatures, which are required for the product. In order to store dried saffron, farmers put it in clean fabric, handkerchiefs or plastic bags, and keep it safe in locked cases or tin boxes inside closets for up to a few months or even a year (13).

Other people put bundles of dried saffron (usually in packets of about 10 grams each) between layers of newspaper and within hard boxes. Stock suppliers and retail sellers also do not have adequate storage facilities for saffron and they usually keep it in large bottles and glass containers with a neck.

Producers also sell the saffron before harvest and a number of them sell it collectively or gradually throughout the year. Most of their products are sold during winter while very little is sold during summer (13).

The method by which middle-men and marketing agents deliver saffron from producers to consumers is seen in the marketing chain. Figure 8-1 shows the chain and the main route through which the maximum supply of saffron reaches the market and the subroutes are shown on either side of this main supply chain. This chain is common to the main saffron-producing cities. Figure 8-1 shows that most dried saffron is being sold to local and city buyers. The latter sell saffron to Tehran, Mashhad and other cities due to their demand. The retailers of these cities prepare their product from 3 sources.

1-from retail sellers production area, 2-from main sellers in Mashhad, Tehran or other large cities, and 3-from packing and processing companies in packed form.

The packing companies also buy massive amounts of saffron from the main sellers in Mashhad, Tehran, and other saffron producing cities. After processing and packing, they export it via saffron exporters. In the marketing chain there are some mediators and sometimes they illegally export saffron to neighboring countries such as Afghanistan, or Pakistan or they may sell it to merchants of different cities and via agents they illegally export it to Europe and US (13).

Based on the results of questionnaires distributed between saffron producers in Torbatehadarieh, Gonabad and Ghaen the contribution of buyers and marketing agents are illustrated in Table 8-1.

On the whole, in the present marketing system, saffron planters, if they had sufficient motivation, would produce premium quality saffron. However, due to difficulties in transportation and distribution, producers receive the least profit. Domestic users of saffron pay high prices for the product, and the maximum benefit

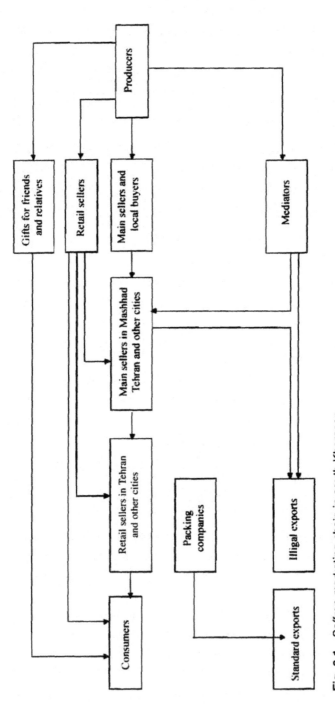

Fig. 8-1 Saffron marketing chain in south Khorasan

Table 8-1 *The contribution (%) of different agents and marketing of saffron in Torbatehaydarieh, Gonabad and Ghaen (13)*

No.	Buyers agents	Torbatehaydarieh	Gonabad	Ghaen
1	Main sellers	73.9	61.1	41.2
2	Retail sellers	8.7	11.1	5.9
3	Mediators	13.1	22.2	47.0
4	Presents	4.3	5.6	5.9

goes to buyers of saffron outside Iran, who sell it at twice the price. (13).

Since saffron producer's act individually the cost of grading, packing and advertising increases and they have to cope with minimum income. Lack of coordination is due to lack of having unity and solid organization concerning buying, selling, distributing, and packing. Because of lack of information, plant producer demand and supply, market situation and fluctuating costs, some people who are not involved in saffron production gain maximum benefit.

Figure 8-2 illustrates that by changing the marketing chain of saffron producers and organizing local unions and establishing packing companies, distribution and exporting saffron through a central point we can prevent the penetration of illegal middlemen and exporters and remove them from the marketing system (13). Therefore, if marketing from production to consumption is undertaken by local buyers of saffron and packing companies, the saffron planter will be receive the benefits. The union buyers of each area are responsible for collecting and buying saffron and also providing some services to their members in saffron producing areas.

8-3 THE PRICE OF SAFFRON

The price of saffron within the country is a function of the export price. Table 8-2 shows that the price of saffron has reached from 25-949 Rials/kg in 1972 to 2654000 Rials/kg in 1999 with a growth rate of 362%. During these 27 years, saffron price has increased 102-fold. This increase in Iran is mainly due to domestic inflation but world wide. The price is reduces from 537 $/kg to 422 $/kg, in spite of huge

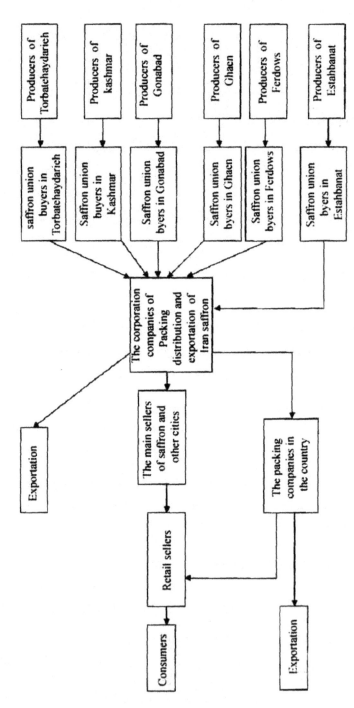

Fig. 8-2 A suggested diagram shows the marketing chain of saffron in Iran. It is noteworthy that union buyers are suggested for cities in which saffron production reaches a minimum of 10 tons per year

Table 8-2   *The per kg price of Iran saffron during 1973 to 2000(7)*

Year	Price Rial/kg	Year	Price Rial/kg*
1973	25949	1987	260000
1974	21700	1988	270000
1975	21678	1989	380000
1976	21786	1990	510000
1977	24217	1991	380000
1978	40166	1992	340000
1979	42032	1993	450000
1980	39776	1994	550000
1981	52145	1995	1300000
1982	70177	1996	1300000
1983	83566	1997	2000000
1984	152985	1998	2236000
1985	252262	1999	2914858
1986	249419	2000	2654000

*in year 2000 the rate was 8000 = 1 US $

increase in production expences. This is the main reason for increasing the saffron acreage. The result of a research (10) indicates that the actual price of saffron did not increase compared to 1973 and in fact, the daily price of saffron is affected by the acreage under saffron. Moreover, a positive and significant relationship exists between saffron exports and acreage, price, and yield/ha of the following year. In recent years the high cost of saffron resulted in a remarkable profit and farmers showed an increased tendency to plant saffron. Unfortunately, after 2002 due to high saffron acreage (50000ha.), the price of saffron declined significcantly, which indicates lack of planning.

8-4   EXPORTING SAFFRON

Saffron is a traditional export crop in Iran, India, Spain and some other Asian and Europian producers. However, due to illegal exporting, exact statistics of its export are not available, but most agree that almost 80% of its production is being exported (12).

Iran with 90% of saffron production is the largest saffron producer in the world. Table 8-3 indicates that exporting saffron

Table 8-3 *The amount (kg) and value ($) of saffron exported from Iran (4)*

Year	Amount (kg)	Price ($/kg)	Total value (x10) (x 10^6)
1971	1100		
1972	-		
1973	400		
1974	900		
1975	900		
1976	700		
1977	2400		
1978	1900		
1979	3084	537	1.66
1980	2017	772	1.56
1981	928	631	0.58
1982	921	775	0.71
1983	601	927	0.56
1984	390	842	0.33
1985	356	561	0.20
1986	670	618	0.41
1987	2711	574	1.56
1988	10420	350	3.66
1989	4963	302	1.5
1990	10860	370	4.02
1991	25411	347	8.82
1992	46150	351	16.2
1993	65924	360	23.73
1994	65101	344	22.40
1995	31600	371.4	11.69
1996	45700	386.6	17.7
1997	36068	390	14.08
1998	59963	399	23.9
1999	82261	414	34.08
2000	106193	410	43.58
2001	121617	422	51.34

from 1.1 ton per year in 1971 increased to 121 ton per year in 2001 which shows an increase of 110 times, and yearly growth rate of 351.6%.

These figures indicate that Iran is the number one saffron producer and exporter in the world, but up to 1988 export of saffron was traditional with low volumes. By using positive enforcement policy of saffron we could observe the acceleration of saffron export. Such increase is not observed in any product, however, some policies such as export exchange and establishing May 1995 convention resulted in reducing saffron export to half of the previous year. Elimination of such rules and regulations during recent years has resulted in encouraging the export of saffron.

During this period the price of saffron was highest in 1983 at $927/kg and lowest in 1994 at $344/kg. The overall trend indicates a high fluctuation in saffron price.

Reports indicate that the world price of saffron exceeds the price that Iran sells saffron to the world market (6). For example, during the years of 1995 and1996, Spain bought saffron from Iran and mixed it with saffron from Spain and with appropriate packing and marketing through large organizations, sold it in the Middle East at a price of $600-1000. The investigation on saffron price indicates that due to lack of a concrete export organization, Iran is not only being damaged by other more competent countries but also suffering a negative effect on the world market price of saffron.

Some expert's (6) suggested that Iran saffron export should be through an organization called the Saffron Export Box (SEM) to control and improve saffron export prices. Producing international saffron enterprises in Khorasan is an example of such a step in recent years. Khazaie (6) recommends that in order to prevent the actual problems of internal and external Iran market and increasing the profit for saffron planters, we need a thorough inspection of the production and marketing. He suggests by using Saffron Box (SB) as a reference under the supervision of members of government and saffron corporation members in order to regulate appropriate policy inspecting and protecting it, we may attain a harmonious system in production, packing, distribution, exports and trouble-shooting of saffron.

Countries importing iranian saffron vary from 14 to 43, and apparently the number of countries interested in buying saffron is increasing during recent years. The main countries buying Iran saffron during recent years include UAE, Spain, Italy, and Germany. Spain which is our saffron competitor, imports Iran saffron in bulk and after premium packing exports it to other countries. Table 8-4 shows the amount of saffron exported to other countries during 2001.

8-5 THE PROBLEM OF EXPORTING SAFFRON

In spite of the importance of saffron and its role in economic and social conditions of saffron producer's areas, several problems exist concerning saffron export. Lack of attention to these problems may result in grave situation for saffron, similar to the market for nuts. The main problems that saffron exports are confronted with are (7).

1. High contamination by microorganisms: considering the safety standards that exist at an international level, mainly in European countries, (the main Iran saffron buyers), infection in saffron is mostly above the standard level and in the long term it may result in a drastic effect on saffron exports.

2. The presence of competitor: recently, countries such as India, China, Morocco, Greece, Turkey and Afghanistan started a program of saffron production, and they may contribute to the world market in the near future.

3. Lack of coordination in production and marketing: development in a planned economic way at different levels of production, processing, packing, handling, marketing, exporting, distribution and consumers coordination and compensation in development should exist for any crop. Presently we do not have such a developed program for saffron. In recent years, due to higher income motivation we did not have such equilibrium between acreage and

Table 8-4   *Saffron export from Iran to different countries during 2001(4)*

Number	Country	weight (kg)	Rials value	Dollars value
1	UAE	56604	4066459160	23170632
2	Spain	41797	31051444905	17693131
3	Italy	7152	5947263270	3388754
4	France	4514	3390837255	1932101
5	Swiss	2252	1842242805	1049714
6	Kuwait	1723	1222373295	696509
7	Japan	1130	932089500	531100
8	Argentina	1100	907335000	517000
9	Taiwan	1266	827973900	471780
10	Sweden	786	730948725	416495
11	Germany	651	544299210	310142
12	Saudi Arabia	673	490050405	279231
13	United States	400	344274740	196168
14	Bahrain	367	278537805	158711
15	England	289	214195995	122049
16	Qatar	285	200745675	114385
17	Finland	105	86609250	49350
18	Singapore	74	61038900	34780
19	Canada	69	57288465	32643
20	China	60	49491000	28200
21	Belgium	39	34510320	19664
22	India	33	27307800	15560
23	Tunisia	30	24745500	14100
24	Oman	29	23638095	13469
25	Australia	28	23415210	13342
26	South Africa	25	21374145	12179
27	Morocco	25	20621250	11750
28	Lebanon	19	18764460	10692
29	Denmark	20	16986645	9679
30	Malaysia	19	15800265	9003
31	Chili	10	8424000	4800
32	Algeria	9	7581600	4320
33	Norway	8	7137585	4067
34	Netherlands	7	5486130	3126
35	Pakistan	5	4212000	2400
36	Philippine	5	4124250	2350
37	New Zealand	3	2527200	1440
38	Shahed Shop	2	2021760	1152
39	Uzbekistan	1	824850	470
40	Cyprus	1	201825	115
41	Austria	1	84240	48
Total	-	121617	90113279490	51346598

production with marketing and export. This situation will be followed by a severe decline in saffron prices.

4. *Saffron adulteration:* Because of high prices, saffron adulteration resulted in a loss in the reputation of saffron from Iran and some other exporters. The most common type of adulteration is staining style with stigma dye or with synthetic colors. Considering the modern analytical and control quality techniques, such adulteration can be identified and result in drastic damage to the export of saffron.

5. *Improper method of harvesting and processing:* Lack of care towards the health of plants and proper technique during flower harvesting, conditioning, handling, transportation, detaching stigma and style, drying, and packing result in lowering saffron quality and a reduction in world demand for saffron.

6. *Improper display of saffron in world markets:* The necessity of exposing saffron standards to acceptable world standard such as ISO Standard.

7. The price of saffron in world market is identified by Spain, whereas, Iran is the main saffron producer of the world and by far produces more saffron than Spain.

8. Illegal export of saffron to Arab countries in bulk results in propaganda against Iran saffron and reduces the price of Iran saffron in the world market.

9. Most people do not know the medicinal porperties of saffron and saffron products.

8-6   SAFFRON INTERNATIONAL STOCK MARKET

The saffron international stock market was established in November 2001 based on item 95 of third development program on merchandize, under the government's responsibility. Since Iran commercial saffron is being sold at a low price in world markets, it would be better to focus on reducing the deficiencies in the current production, packaging, transportation and marketing of saffrom rather than trying to add value to the product. It is also important to prevent adulteration in the world market. The Saffron Stock Market would then benefit the saffron producers, buyers and exporters. (8).

The stock market could complement the activities of other responsible organizations involved in promoting saffron quality, by emphasizing quality standards, and actual price. For example, in recent years the ministry of Jahad Agriculture have started organizing workshops and distributing sieves (screens), heater and other extension activities to reduce contamination and promote saffron quality. However, practically in traditional markets there is no difference between the traditional product and one of a higher quality, and farmers do not have any motivation to present a standard product. In a stock market this difference is taken into account while determining the price, and the farmer has to produce a certain quality. Since we can–not sort saffron like other products, the value is dependent on packing and storing saffron. Accepting the external demand for quality and packing will drive the producers toward consumer desire (8). In stock markets, farmers may sell their products either in cash or on advance payment. Nowadays, trading is immediate and in cash.

The Mashhad International Stock Market has prepared the groundwork by establishing delegates to purchase saffron from farmers, Bureau of Saffron Commerce, pricing committee, quality and standard measurement commerce hall and dealing inspector, Bureau of Judgment and Storage rooms.

Establishing these units produces a hot market for saffron. Information is broadcast internally via telephone and other media and externally mostly via the Internet site www.mashade.com. The saffron purchaser may buy any amount of saffron from Saffron Stock Market and will be sure of the quality and price.

Activities in stock market are open to everybody and no obligation exists in selling, buying or dealing and any complaint could be prosecuted. The policies the stock market follows includes the following (1).

1. Publishing regular brochures about Iran saffron in Persian, English, and Arabic

2. Publishing a book about saffron

3. Cooperation of stock market employees including unions, corporations, and main saffron exporters in Iran Commerce Room in Dubai

4. Participation of saffron exporters in exhibitions overseas at least once a year

5. Establishing Research and Development Department by keeping an eye on saffron market and marketing with cooperation of universities and research and technology parks

6. Reduction or tax-free revenue, because this is frustrating for employees and results in deal masking, especially because in common markets saffron benefits do not exceed more than 2-3% the attempt in reduction or compensating tax payment or method of obtaining tax from Wall Street related to agricultural firms could be important. It is noteworthy that the first product made by hand, which in being sold legally, does not need tax payment and exports are also free from payment of tax

7. Enough cash payment to buy saffron around the year and prevent severe cash shortages to save saffron from negative competition of saffron exporters. To establish a true and rational price for saffron in the production centers.

SUMMARY

Plantation and production of saffron is an old traditional practice. Saffron requires little water, and since most of Iran and especially the Khorasan region have low availability of water, this is an ideal crop. The main body of saffron production is exported and since Iran is undoubtedly the main producer and exporter of saffron, developing saffron production and exporting is important to enhance farmers income and introduce exchange to the country. Considering the economic, social, historical, and exceptional position of Iran saffron in the world, saffron needs undisputed support in the processes of production and distribution and export via policy makers in the Department of Agriculture and Government. The current system did not provide support between acreage development and production with marketing and export systems. Saffron was not distributed through a proper channel and marketing system. However, some steps such as establishing saffron corporations and stock markets is promising, but their activities requires appropriate investments in production, distribution, export and coordination among these different activities. The authorities and policy makers also need to pay special attention to other agricultural crops.

References

1. Agricultural Jahad Ministry. Agricultural statistics. Office of Statistic and Information (Technical Reports).
2. Dehghanian, S. and M. R. Kohansal. 1997. Pricing policy and effect of saffron importation on saffron acreage, price, and yield per hectare. Journal of Agricultural Economics and Developments, number 14.
3. Hossaini, M. 1998. Investigation and evaluation of economical and sociological effects of 10 years work on saffron. Iran Industrial and Scientific Research Organization, Khorasan Center (Technical report).
4. Iran Custom Compendium (several years).
5. Kamijani, A. Rules and regulation of accessing agricultural products in general trade agreements and its effect on Iran agricultural economics.
6. Khazaei, A. 1998. The study of saffron fluctuation, necessity of producing corporations. Journal of Agricultural Economics and Developments, number 19.
7. Khorasan Agricultural Jahad Organization. 2002. Internal reports on programming and design managements.
8. Mashhad Saffron International Stock Market. 2003. Internal Technical Report.

9. Mohammadi, F. 1977. The study of saffron and cumin production and exportation. Agricultural Economics and Developments, Seminar on selected articles.

10. Research, Education, and Extension Organization. 1997. Iran saffron techno- economics production study (Technical report).

11. SAbzavari. 1996. Saffron, the red gold of Kavir. Agricultural Bank Publication. Number 46 (Technical report).

12. Sadeghi, B. 1995. Saffron, the economical and sociological effect, and the necessity of evolution in its tradeship. Second conference on saffron and medicinal plant cultivations. Gonabad.

13. Turkamani, J. 2001. Production, marketing, and factors affecting on saffron importation. Agricultural Economics and Research Programming Institute (Technical report).

CHAPTER 9

Role of Indigenous Knowledge in Traditional Agriculture with Emphasis on Saffron

A. Koocheki
Faculty of Agriculture, Ferdowsi University of Mashhad, Iran.

9-1 INTRODUCTION

Indigenous knowledge is an important cultural and technological aspect of a community which develops through careful observation and experiences of the surrounding natural environment by local people. This type of knowledge is unique for a society or cultural group and is an indication of their innovation and know how. Local technologies are the basis of decision making in agriculture, health, natural resource management and other activities, and is embedded in community practices, institutions, relationships and rituals.

Although indigenous knowledge has a long history in human societies, the necessity for its utilization in development is rather new. In the last two decades use of indigenous knowledge in sustainable development has been widely recognized.

From the Second World War, development processes have faced several stages. In the first stage emphasis was focused on economic growth, then equity was a concern, followed by the requirements and needs of human societies, participatory development and finally sustainable development was the main issue. Agricultural development was not an exception in this process.

After the Second World War, development in agriculture was on the basis of transfer of science and technology from the developed to the developing nations and the extension agents were responsible for a widespread expansion of these technologies. Since most farmers in these countries were not able to adopt such technologies they resisted these changes and such resistance was considered backwardness and an old fashioned attitude. One of the examples of relative failure of transfer of technology to the poor farmers was the Green Revolution of the 1960s. This technology, based on high inputs, was initially supported by developed nations, but in the longer term farmers could not afford these inputs and therefore it was not widely accepted (1).

Today new concepts have evolved in utilization of indigenous knowledge in agricultural development in developing countries. However, this type of knowledge which has been developed on the basis of trial and error in a particular society is different in nature, content and methodology from conventional scientific knowledge which is fundamentally based on experiences and trials carried out by scientific institutions. Scientific knowledge has universal scales and it is experienced and observed and the discovery, documentation, records and training of this type of knowledge is different from indigenous knowledge. For the latter there is no honoring mechanism for the producers and no intellectual property right is observed, while for the former there is a distinct honoring system for the discoverer and strict property rights are enforced.

Indigenous knowledge has an economic value unlike scientific knowledge which has no trading value. This type of knowledge is based on cheap, low input approaches and since they have been developed over a long time in response to the natural environment of the local societies, it has a strong ecological nature and participatory basis in contrast to the scientific knowledge which is based on individuals, objectives and technologically based attitude (6, 5).

Indigenous knowledge (IK) belongs to all individuals in the society and women play a significant role in the development of IK (15). As an example, in developing countries, women contribute a great deal in agricultural activities. Nomadic women are involved in dairy processing and handicraft production exclusively. In Iran these groups contribute 60% in rice production, 90% in vegetable production, 30% in fruits and 90% in silk worm raising (14). Women are also involved in conservation of biological diversity and in particular agricultural biodiversity and also in identification, classification and utilization of medicinal plants.

9-2 TRADITIONAL AGRICULTURE AND IK

Production in traditional agriculture is based on long term sustainability, rather than short term productivity. Sustaining productivity of marginal and poor lands, enhancing biodiversity which reduces the economic risk; and spatial and temporal distribution of resources and, therefore, proper allocation of labor force through the year are fundamentals of traditional agriculture. Most traditional ecosystems are at the center of the origin of agricultural crops. Traditional farmers utilize the microclimates which are very diverse in terms of soil, water, temperature, altitude, land slope and fertility efficiently.

It has been estimated that 25,000 to 75,000 species of medicinal plants are utilized in traditional medicine, of which only one percent has been identified by scientists and is in commercial use. The modern pharmaceutical industry is still very dependent on plants discovered by the local communities and to date the annual trade value of up to 7,000 medicinal plant constituents is estimated to be 35 to 47 billion dollars (2). Nearly 25% of the medicines used in the USA are provided from the developing countries (6).

At present, most of the problems associated with agriculture and food production in the third world is unsolved due to the nature of these problems and the structure of the societies in these countries, which as very different from those of the developed nations.

Diversity of food production systems in developing countries is in contrast to the high input food production technologies of the developed nations and, therefore, many local crops have been substituted by the modern improved varieties with a high demand of inputs. Research funds are mostly allocated to plants adapted to high input and new technologies and a growing number of scientific journals are published on issues related to these crops.

9-3 INDIGENOUS KNOWLEDGE IN IRANIAN AGRICULTURE

It has been seen that Iran is one of the main centers of plant and animal domestication since the last 10,000 years (8). Based on this historical background many indigenous technologies associated with food production have evolved in this part of the world. Underground water extraction "qanat" with a history of 3,000 years was developed in Iran (12). Qanat with the surrounding cultural environment has been one of the strongest socio-economic structures of agricultural communities of the area. This technology, which was the most advanced of its time, spread from here to distant parts of the world and has now been replaced by the modern pumping system. Crop production in water scarce areas based on a highly efficient use of water through proper utilization of planting pattern and use of low demanding crops was developed in conjunction with the qanat. This is still in operation in the southern parts of Khorasan and cultivation of saffron, cumin and barberry which all are low water demanding plants is a good example.

Nomadic system of food production which is based on efficient use of feed sources through seasonal immigration is another example of IK, not only in food production but also in associated activities such as handicraft and carpet production.

9-4 EVALUATION OF IK IN SAFFRON PRODUCTION

Saffron production technologies have not changed much through history. This is mainly due to the nature of this crop which is a high labor demanding and family oriented plant. Therefore not much substitution has been made in terns of production, processing and utilization. This delicate crop has tremendous economic, social and cultural values for the local farming communities of southern Khorasan in Iran and Kashmir in India and has been a source of income for small low income farmers.

Saffron production systems can be viewed in terms of technology, socio-economic and cultural aspects. In terms of technology, although saffron is a low water demanding plant and its production cycle and vegetative growth coincides with autumn and winter, primary irrigations and particularly the first one is very crucial to timeliness of flower emergence and, therefore, availability of water at the beginning of growing season is very important. Qanat was, and still to some extent, is the main source of water for saffron in Khorasan. Most saffron farmers are small holders and share water from the qanat ranges. Therefore, water use and allocation was based on a very precise calculation and measurement. This measurement was made on the basis of scales called "bowls", "cups" and "drops" which indicate scarcity of water in the area.

Proximity of saffron to qanat outlets located near the villages was important for two main reasons, the first being easy access for harvesting and transferring flowers early in the morning and second, facilitation of the primary irrigation. Since this was also applicable for orchards and gardens, combination of saffron fields and orchards in an integrative way gave a pleasant view to the villages and surrounding landscape. Therefore, qanat and the culture that grew up around were integrated with the cultural integrity of saffron and this provided a rhythm to life in the villages (10, 12). Nutrient requirements of saffron is normally provided by animal manure and in the old days when saffron production was also practiced in the central parts of the country, part of the manure requirement was collected from pigeon towers, artificial structures built for this purpose. In each tower thousands of pigeonholes were made in such a way that the feaces would drop in the center of the floor for further collection. The earliest of these structures is estimated to have been built more than a thousand years ago (7).

It is worth mentioning that, even at the present day, based on the nature of saffron crop which still receives most of it nutrient requirements from animal manure, there are not many pests and diseases that have to be combated with chemicals, and saffron could be regarded as an "organic produce", though not a certified one based on the present day certification standards (11) rather could be called "organic by neglect". In terms of social, economic and cultural values, saffron production was a family practice, which was arranged in such a way that when extra labor was needed the whole community was involved in production and processing based on sharing or exchange of labor.

Saffron has been a constant ingredient of many Iranian cuisines and inclusion of saffron in foods has been regarded as prestigious. During feasts, festivals and religious ceremonies, dishes were decorated and flavored by saffron. Saffron was and still is a valuable gift and in some cases was included in the dowry of newly married couples. Saffron was also used in paintings, miniatures and textile dying. It has also been used in verdigris as an inhibitor in Persian miniature paintings and also for preservation of old manuscripts (4).

In an attempt to evaluate IK associated with saffron production and utilization in southern Khorasan, a survey was made of 25 farms in the five main growing areas of saffron, Torbate hydariah (4 farms), Ghaen (10 farms), Sadeh (3 farms), Kazar (3 farms) and Birjand (8 farms). For this purpose interviews and discussions were held with elderly local inhabitants and a questionnaire containing different information on production practices, processing and utilization purposes was developed. Based on the information appropriate statistical analysis was carried out and Fidelity Level (FL) or the percentage of informants claiming use of saffron for the main purposes was calculated. Since some uses which received high FL values might have been known to only a fraction of the participants, an appropriate correction factor was calculated on the basis of the relationship that existed between FL of a particular type of use with that with the highest FL. The corrected fidelity level, or Rank Order priority (ROP) for a given type of use was then calculated by: ROP=FL×RPL. Therefore, ROP was used to classify different type of uses for saffron (9). Results of this investigation were presented in the First International Symposium on Saffron Biology and Biotechnology held from 22-25 October 2003 in Albacete, Spain (13).

In Table 9-1 number of respondents from each area for each category of questions is presented.

In Table 9-2 coefficient of variation for farming practices and type of uses is shown.

As seen, type of uses was diverse which indicates that there is a high variability between different areas is terms of saffron utilization. In general, farming practices were also varied amongst the area but in cases such as irrigation or methods of planting, there are not many differences between the areas.

- *Selection of corm size:* More than 90% of farmers use their own corms for selection of corm size, 68% of farmers use large corms and this promotes stand establishment in the first year. However, medium size corms were preferred by 17.8% of farmers. Sadeghi (16) reported that corms with more than 5 grams could produce flowers in the first year.
- *Irrigation:* 98% of farmers consider the first irrigation after summer dormancy as the most crucial one for timeliness of flower emergence. Further irrigation depends on the availability of soil water and varies from 10 to 30 days interval. However, Alizadeh (3) reported that a 15-day interval gave better results than longer periods of 30 or 60 days. He concluded that water requirement for saffron is 300 mm or 3000 m^3/ha annually.
- *Methods of planting:* Hill planting and row planting are the main planting methods and the number of corms used per hills varies considerably from 5-15 corms.
- *Manure and fertilizer application:* Animal manure is the most common practice for soil fertility enhancement and chemical

Table 9-1 *Number of respondents in each area*

Questions	Responses	T	G	S	K	B	Total
Traditional methods of use	1- as food coloring and spice	4	5	2	2	3	16
	2- as hot medicine	2	7	3	1	5	18
	3- as medicine for bone remedy	1	3	1	1	1	7
	4- additive in tea	1	3	2	1	2	9
	5- additive in bread	0	2	2	0	0	4
Corm selection for planting	1- use of large corms	3	7	2	2	5	19
	2- use of radium corms	2	1	1	1	0	5
	3- use of corms grown in clay soil with fresh water	0	3	1	0	0	4
	4- use of corms from other areas	1	0	0	0	2	3
Methods of planting	1- row hill planting	3	3	1	2	5	14
	2- hill planting	1	4	2	1	3	11
	3- planting 4-5 corms/hill	3	5	0	0	5	13
	4- planting 10-15 corms/hill	0	0	2	3	4	9
Methods of irrigation	1- one irrigation for breaking the crust	4	8	3	4	8	27
	2- Irrigating every 10-15 days	1	6	2	2	4	15
	3- Irrigating every 30 days	0	4	0	0	3	7
	4- No winter irrigation to avoid freezing	0	2	2	0	0	4
	5- Use of seasonal floods	0	1	2	0	0	3
Manure and chemical fertilizer application	1- use of manure	3	9	3	2	8	25
	2- use of chemical fertilizer	3	2	2	2	3	12
	3- use of aqua as fertilizers and micro nutrients	1	0	0	1	0	2
	4- sheep manure through grazing on the field	0	0	2	0	0	2
Protection of the field	1- timely irrigation	2	4	1	0	0	7
	2- weed control	3	5	2	2	8	15

(Table 9.1 Contd.)

(*Table 9.1 Contd.*)

		T	G	S	K	B	
	3- mouse and rabbit control	2	3	1	3	5	14
	4- avoiding animal entry in the field	0	6	2	1	4	13
Picking flowers and separation	1- picking before sunrise	4	9	3	3	8	27
	2- separating the whole stigma plus style (Poshal)	2	4	2	2	5	15
	3- separating the stigma only	1	2	1	1	3	8
	4- separating on the same day as picking flowers	0	2	1	0	0	13
Storing saffron	1- storing in packed metal containers	3	5	2	2	5	17
	2- storing in white cloth	1	3	1	1	2	8
	3- storing in non-plastic containers	1	3	0	0	2	6
	4- storing in non-paper containers	0	2	0	0	0	2
	5- storing in wooden containers	0	0	1	0	0	1
Use of plant parts	1- harvesting the leaves for animal feed	3	6	3	3	7	22
	2- direct grazing	1	2	0	0	1	4
	3- burning the residue if infested heavily with weeds	0	0	1	0	0	1
Use of other pats of saffron	1- as animal feed	2	3	0	1	2	8
	2- dying clothes	1	4	1	2	4	12
	3- medicinal use	1	1	0	0	1	3

T = Torbat Hydariah, G = Ghaen, S = Sadeh, K = Kazar B = Birjand

fertilizers are used only by 43% of farmers. Other practices were not common.

- *Farm protection:* There are not many serious pests and diseases for saffron. However, mouse and other rodents such as rabbit are a serious threat to the corms. Weed control and preventing animals entering the fields are other means for protecting saffron fields.

Table 9-2 *Coefficient of variation for farming practices and type of uses of saffron in southern Khorasan, Iran (%)*

Practices and uses	T	G	S	K	B	Whole area
Type of uses	41	68	31	18	22	38
Corm selection	35	37	32	19	20	31
Methods of planting	12	16	17	21	33	34
Irrigation	15	52	41	14	18	21
Manure and fertilizer application	21	24	29	38	22	19
Field protection	32	38	30	27	28	38
Picking flowers	42	44	27	11	30	23
Use of plant parts	38	31	27	24	35	27

- *Picking flowers:* more than 95% of farmers prefer picking flowers before sunrise and all farmers picked flowers by hand. Normally flowering period is between 10 and 20 days. Separation of saffron from the flowers depends on the goal of production and it may be the whole style and stigma kept loosely which is called Poshal, the same part kept in bunches is called Dastah and only the stigma is called Sargol. These are kept in different containers but the most common containers are tight wooden or glass jars.

- *Type of uses:* In Table 9-3 types of uses are presented, and as is seen, most of the respondents use saffron as a hot medicine and other uses.

In descending order are: food color and spice, additive in tea, medicine for bone remedy and additive in bread. However, new findings show that saffron has other noble uses such as cancer cure and other medicinal purposes

Table 9-3 *Type of uses of saffron in southern Khorasan, Iran.*

Type of uses	Fidelity	RPL	ROP
Food color and spice	57	88.8	50.7(2)*
Hot medicine	64	100	64.2(1)
Medicine for bone remedy	36	38.9	9.7(4)
Additive in tea	32	50	16(3)
Additive in bread	14	22	3(5)

*Figures in parenthesis are ranked as per use

- Use of other parts: More than 75% of farmers use saffron leaves as hay for animal feed and 14% prefer direct grazing. Valizadeh (17) reported satisfactory results for the feeding value of leaves for animal grazing. Burning the residue is not a common practice. Other parts are also utilized as coloring agents, medicine and as animal feed.

SUMMARY

Indigenous knowledge in Saffron production, processing and use is diverse and should be recognized, documented and if necessary modified to new technologies. However, it must be born in mind that traditional knowledge like any other phenomenon is in a constant state of change and as each generation matures, skills perceived as immediately useful are gained while other with lesser perception of immediate value may be lost. Now indigenous knowledge is in danger of disappearing not only under the influence of global processes of rapid change, but also because the capacity and facilities needed to document, evaluate, validate, protect and disseminate such knowledge are lacking in developing countries. Since saffron is an important crop in Iran in terms of economic, social and cultural values and Iran ranks first in the world in terms of acreage, production and export to the international markets, it is therefore our duty to document and protect indigenous knowledge associated with saffron production, processing and use in the country for further reference. However, more research needs to be done and methods need to be developed on collection, documentation, validation and evaluation of this type of knowledge.

Acknowledgment

Collaboration of Dr. M. Nassiri and Mr. M. Behdani in the survey is gratefully appreciated.

References

1. Agrawal, A. 1995. Indigenous and scientific knowledge: some critical comments. Indigenous Knowledge and Development Monitor, Vol. 3. Issue 3: 3-6.

2. Aguilar, Q. 2001. Access to genetic resources and protection of traditional knowledge in the territories of indigenous peoples. Environmental Science and Policy 4: 241-256.

3. Alizadeh, A. 2003. P. 122-135. In: Kafi, M., M. H. Rashed, A. Koocheki and A. Mollafilabi (eds). Saffron: Production, Technology and Processing. Center of Excellence for Special Crops, Faculty of Agriculture, Ferdowsi University of Mashhad, Iran.

4. Barkeshli, M. and G. H. Ataie. 2002. pH stability of Saffron used in verdigris as an inhibitor in Persian miniature paintings. Restaurator. 71: 154-164.

5. Cashman, K. 1989. Agricultural research centers and indigenous knowledge systems in a worldwide perspective: when do we go from? In: Michael, W., L. J. Slikkerveer and S. O. Titolala: Indigenous knowledge systems. Implication for agriculture and international development. Studies in Technology and Social Change, No. 11. Technology and social change Program. Iowa State University, Amer, Iowa.

6. Emadi, A. and A. Esfandian. 2000. Application of indigenous knowledge in sustainable development. Vol. 1. Principles, concepts and beliefs. Ministry of Agriculture, Research Center for Rural Studies.

7. Farhadi, M. 1993. Survey on the history of pigeon house in Iran. Publication and video center of Jehade Sazandaghi, Tehran, Iran.

8. Flanner, K. V. 1962. In : Ueko, P. J. and B. W. Dimbleby (eds). The domestication and exploitation of plants and animals. Gerald and Duckworth Co. London.

9. Friedman, J., Z. Yaniv, A. Dafni and D. Palewitch. 1986. A preliminary classification of the healing potential of medicinal plants based on a rational analysis of an ethnopharmacological field survey among Bedouins in the Negev desert. Israel J. Ethnopharmacol. 16: 275-287.

10. Honari, M. 1979. Qanats and human ecology. Ph.D Thesis, University of Edinburgh, Scotland.

11. Koocheki, A. 1995. Organic Saffron in Iran. Ecol. Farm, 10: 8.

12. Koocheki, A. 1996. Qanat, a sustainable ancient system for exploitation of underground water in Iran. Proc. 11th International Organic Agriculture Conference. 12-15 August 1996, Denmark.

13. Koocheki, A. 2003. Indigenous knowledge in agriculture, particular reference to saffron production in Iran. Proceedings of the First International Symposium on Saffron Biology and Biotechnology., 22-25 Oct 2003, Albacete, Spain.

14. Mahsaii Zadeh, A. 1996. Sociological evaluation of the role of women in Iranian agriculture. Quarterly Journal of Agriculture and Development, special issue on role of women in agriculture. pp. 64-80.

15. Roth, G. 2001. The position of farmer's local knowledge within agricultural extension, research and development cooperation. Indigenous knowledge and Development Monitor, Vol. 9. Issues 10-12.

16. Sadeghi, B. 1980. Effects of corm size on flower production in Saffron. Scientific and Industrial Research Organization, Khorasan Center, Annual report.

17. Valizadeh, R. 1988. Use of Saffron leaves as animal feed. Scientific and Industrial Research Organization, Khorasan Center, Annual report.

CHAPTER 10

Processing, Chemical Composition and the Standards of Saffron

A. Hemmati Kakhki
Iran Scientific and Industrial Research Institution, Khorasan Center

10-1 INTRODUCTION

Saffron, since the ancient times, has been in used to flavor different kinds of dishes, because of its beautiful color, aroma and exceptional taste. These days, people again prefer to use the natural products rather than the chemical and synthetic ones. Saffron is used for getting natural color, as a medicinal herb to reduce level of bilirubin of the blood and anti-cancer medicines (5). In the same manner, the petal of saffron can be used as means to give food color as an anti-syanin (21). Its leaves are also used as fodder for domestic animals (46).

Iran is considered to be the biggest producer and exporter of saffron in the world (27). The production and presentation of saffron with a desirable quality, the base requirement, the expectation of the customer and the market can be a source of expanding the trade of this plant and placing it on a firm footing. For this purpose, it is necessary to have knowledge about the quality over production, recognition of its peculiarities, the compositions of saffron, the correct processes and different stages to reach this product, information about the established standards recognizing its properties and the use and consumption.

10-2 THE PROCESSING OF SAFFRON

The word processing of saffron applies to all stages after flower harvesting in order to deliver an acceptable product to the market. This work is done either by the producers themselves or by specialized workshops and factories. The object of this is to bring it to the market loaded with the quality suitable to the taste of the consumer. In the processing of saffron, it may happen that the duration between processing to consumption may be long. This particular period is very important from the point of view of maintaining the quality of the product. The processing of saffron passes through several stages or procedures. These are, picking the flowers, transportation of flowers, separation of stigma, sampling, testing, sorting, packing and bringing saffron to the market for sale. There are also other matters related to the capability of saffron in its full form that may be effective in its extraction and presenting it to those who demand for it. This will be dealt in this chapter in a separate section that would deal with its peculiar combinations.

Figure 10-1 shows the different stages of the processing of saffron, right from picking of the flower to distribution and consumption of the product (24):

10-2-1 Harvesting, Transporting and the Preservation of Saffron Flower

2-3 weeks after the first irrigation of the field, the saffron flowers start to appear, but this is conditional to the temperature. The time of flowering in the saffron producing regions of Khorasan province starts by the 15th of October and continues till the 10th of November. This is also bound to the regional conditions. Usually the

Saffron flower

↓

Harvesting

↓

Separation of stigma

↓

Drying process of saffron

↓

Maintenance of saffron

↓

Transporting saffron to packing plants

↓

Sampling and testing

↓

Weighing

↓

Packing

↓

Sampling and testing of the final product

↓

Sale

Fig. 10-1   Different stages of processing saffron from flower harvesting to consumption.

process of flowering and picking takes place between 15 and 20 days. During this period the picking starts early in the morning every day (22).

The national standard of Iran (No. 5230) under the rule of harvesting and processing of saffron has given the following recommendations for picking flowers, which are applicable before packing:

The time of harvesting

It is proper that the time of picking flowers should begin in the early morning hours (temperature in the morning is low). As the picking of flowers is easier when still in the bud, it, therefore, reduces the mechanical damages to the flower during transportation.

Flower pickers

The labourers who pluck flowers should be healthy, particularly their hands should be clean and they must not have any skin disease or infection.

Flower container

For the purpose of transporting plucked flowers either straw baskets or clean and dry plastic buckets should be used.

Transportation

The transportation of flowers of saffron should be done in such a manner that it may not suffer any damage and it should be kept safe from pollutants. The piling should not be beyond the limit and flowers should not be pressed while transporting. After transportation, up to the time of separating the stigmas from other parts of the flower, the flowers should be kept at a clean and cool place, far away from sunlight (22).

The quantity of the flowers that can be harvested from the unit surface area of a farm depends on many factors, such as, the age of the field, the management of the field and the climatic factors. The quantity of flower yield in the first year after planting is minimal, however, in the subsequent years the yield gradually grows and then starts to decline. Table 10-1 shows the quantity of flowers and dried saffron, collected from one hectare in different years of the age of the field (3).

Information about the other parts of the flower by measuring with the scale could be useful for estimation of economic yield.

Table 10-1   *The yield of saffron in different years of age of the field (3)*

Age of the field (years)	Flowers yield (kg/hectare)	Dry saffron yield (kg/hectare)
1st Year	100	1.32
2nd Year	400	3.95
3rd Year	600	7.80
4th Year	800	10.50
5th Year	1000	13.20
6th Year	700	9.15
7th Year	600	7.80
8th Year	400	5.20
Average of 8 years	562.5	7.36

In an experiment to evaluate the other parts of the flower, nine fields of varied age of Gunaabad town were selected. In these fields repeated tests were performed four times. In each repetition, by coincidence, 150 grams of flowers were gathered. After that the balance of the parts of the flower was calculated by separating them from the flower (29).

The results of these tests showed the varied age of the fields had no effect in the number of flowers in a unit weight. On average in each kilogram of fresh flower there were 2173 flowers (with the coefficient of variation = 9.73%). Table 10-2 shows the weight of the parts of the flower on the basis when they were fresh or dry.

For producing one kilogram of dry stigma (superior saffron) 105.4 kg of fresh flowers are required, which is equal to 230,000 flowers. For producing one kg of dry stigma and style (standard saffron) 78.5 kg of fresh flowers are required, which is equal to 170,000 flowers. By taking these results into consideration, it becomes quite clear that in each 100 g standard saffron there is 25.5 g of style and 74.5 g of stigma. In the same manner the weight of the straw of the dry saffron from each kg of fresh flowers is about 9.82 g. This is conditional to the length of the style, which along with the stigma should not be more than 5 mm. Therefore, for producing one kg of high quality saffron, 101.8 kg of fresh flowers will be needed, that would be

Table 10-2 *The average weight of the parts of the flower that comprised one kilogram of fresh flower (29)*

	Stigma (g)	Style (g)	Stamen (g)	Sepal and Petal (g)
Fresh weight	47.39	28.93	59.35	864.33
Dry weight	9.48	3.26	14.78	98.36

221,000 flowers. Taking into account the effect of different regional conditions, physical and chemical properties of the soil, particularly the time of harvesting the flowers, the above results can vary in exceptional cases up to 10% (29).

The tests that have been done (38) on saffron in Kashmir, India, show that the parts of a fresh saffron flower contain 8% stigma, 2% style, 80% sepals and petals, 8% stamen and 2% residue. The weight of one three-branched stigma in that region has been reported from 16–27 mg, the length of the stigma from 18–35 mm and the diameter from 3–4 mm.

The time of harvesting the flowers from the field is the foremost factor that has a great effect on the quality and quantity of saffron. Raina and his associates have done research effect of the time of harvesting of the flowers in the processing of saffron and the color strength in Jammu and Kashmir. These scholars tested the flowers at three stages. They took the balance weight of the product in dried form and examined its traits in all the three stages. The result (Table 10-3) showed that when the saffron is taken in the form of developed flower (of 5 days) its yield gets increased as compared to the bud. In that case the yield increases by 42% and the color strength increases by 7%.

10-2-2 The Separation of Stigma

When flowers are taken from the field the separation work of the stigma from the other parts of the flower should start as early as possible otherwise the rapid decay of the flower makes it useless. The place where the separation work is done must be clean and hygienic.

Table 10-3   *The saffron stigma specification at different stages of flower development (38)*

Flower stages	Average weight of stigma[1] (g)	Average diameter of stigma[1] (mm)	Average length of stigma[1] (mm)	Crosine in fresh stigma[2] (g/kg)	Crosine in all other parts of the flower (g/kg)
Bud as on the 1st day of emergence	0.026	3.0	28.0	165.4	0
Flower 3 days after emergence	0.032	3.6	31.5	167.2	0
Flower 5 days after emergence	0.033	4.0	35.0	170.7	0

[1]The average of one cycle of three years from different regions.
[2]On the basis of dry weight.

People involved in the separation work should be aware of their health problems.

The separation work in Iran is done in three forms. In the first method, the three-branched stigma along with the style is separated from the other parts of the flower. They are placed in order on each other. They are dried in the same manner. The saffron prepared by these means is technically called Zafran-e-Dasta (handled saffron).

In the second method, when the flowers are plucked, the style and stigma are separated. The stigma is cut at the point where it is joined with the style. Each one is collected separately and dried. The dried stigma collected by this method is technically called Zafran-e-Sar-Gol (saffron from the top of the flower). However, at the time when the bunched saffron is being transported, some amount of the stigma gets separated from the joining point with the style. Some producers separate the stigma from the style in the form of bunch. They do this when the saffron is dry.

In the third method, usually the three-branched stigma along with a small part of the style gets separated from the other parts of the flower. This is placed in a clean vessel to dry. The length of each style along with the stigma, in this method, depends upon the quality of saffron and the requirement of the customer, that varies from 1 mm to almost 10 mms and sometimes more than that. The saffron prepared by this process is technically termed as Zafran-e-Pushali (straw saffron) (11). Table 10-3 in supplement section shows different bunches of saffron, the top of the flower and the straw covered (Pushali).

The National Standard of Saffron (22) (this rule applies to saffron after harvesting and processing but before placing) has given its recommendations regarding the process of separating. They have suggested production of fibril from saffron with each three-branched stigma and it should be with maximum 11mm style and placed in loose form in a clean vessel to make it dry immediately. Separation of stigma in the form of bunches prepare the conditions of growth, reproduction of microorganisms, increase in pollution, reducing the quality of product and increased time for drying. For this process the following advantages have been mentioned:

- The reduction in the possibility of pollutants
- Fast process of separation
- The increase of the apparent volume of the product
- Preserving the quality of saffron
- Reducing drying time (22)

To keep saffron preserved for a longer period of time, it should be dried. The process of drying has great effect on the quality and the worth of the final product. In different countries and regions saffron is dried in different ways (3, 10, 11). In the traditional Iranian system when the stigma and style are being separated from the flower, it is spread in rows either on cloth or a piece of paper and is dried in shade. This system is not free from defects. The main defect is that it takes a longer time to dry. The longer span of time increases the probability of microorganism growth and also pollution. In the last few years this system is gradually losing ground and new methods are employed.

In Spain, silken net sieves of 30 cm diameter are being used. The fresh stigma is placed in layers of 2-3 cm thickness on the net. After that the sieve is kept at a proper distance before a heating system. The sieves are placed one upon the other in rows and their location is changed. With this process the product dries uniformly and evenly. With the purpose of maintaining the quality of saffron by shortening the drying time by this simple means, this system is gaining popularity over the traditional systems (26).

The other system of drying saffron is electric ovens. The temperature of these ovens can be regulated from 50°C to 60°C and are equipped with particular trays on which the silken nets are placed. In this system the saffron is kept in layers of 1-2 cm thickness for a period of 30-40 minutes and the saffron dries with the heat.

In India saffron is dried by two methods. In the first method the stigma and style are separated from the other parts of the flower and are then placed directly under sunlight. The sunlight takes away the moisture till the water content of stigma reaches 10-12%. In accordance with the temperature it takes 3-5 days to dry the product. In the second method the whole flower is dried under sunlight rather than drying the stigma and style. After that the stigma is separated from the flower by hand.

With the purpose of evaluating the effects of different systems of drying saffron, Raina and his associates (38) dried saffron using different means. After this they evaluated quality of color, aroma and smell of saffron. Table 10-4 and 10-5 indicates the results of their research.

In these experiments the total pigment concentration in the samples varied from 11-17% of the weight of the dry saffron. The examination of the results showed that in all methods of drying that the temperature was less than 30°C., time taken to dry saffron has been longer (27 to 35 hours) and the final quality reduced. This happened, probably due to the longer activity of enzymes that causes the analysis of crocine. Drying saffron at the abovementioned temperature (more than 60°C), takes shorter time (between 2 to 4

Table 10-4   *The effect of different systems of drying on the level of the color and the time of drying saffron (38)*

Drying method	Temperature (Centigrade)	Quantity of Crocine[1] (mg/g)	Drying time-1 (Hours)
1. Drying in shade.	4-18	110.0	53.0
2. Drying in sunlight	15-21	127.0	28.0
3. Drying in sundryer.	35-49	153.0	6.5
4. Drying in electric oven.	*(Temperature of the incoming air)*		
	20	124.0	22.0
	40	154.6	5.3
	50	148.5	3.8
	65	134.0	3.0
	80	115.5	2.4
5. Drying by blowing air from the opposite direction.	20	129.4	25.0
	40	156.4	7.0
	50	150.4	4.5
6. Drying in an oven under vacuum. *(40 mm mercury)*	40	170.5	5.0
	50	161.0	3.6
	65	144.7	3.0
7. Drying with a desiccators	40	156.7	4.6

[1]This testing has been done successively and the samples have been collected from different regions of Kashmir, India.

hours). However, the thermal analysis of the pigments could result in reducing the quality of the product. The samples of saffron that have been dried at temperatures between 35 and 50°C in the electric oven, vacuum oven, sun or by blowing air from the opposite direction, take about 6.5 hours to dry and the color strength is similar to that of the fresh saffron. Besides the samples were good in texture, good color and appearance was quite bright.

On the whole, the stigma of fresh saffron does not have any smell. These compositions take place by means of thermal and picrocrocine enzymatic analysis [4-beta-(D-Gluco Pironozyloxy) –2-, 6, 6-Tri-methyl –1-Cyclohexon-1-Carbo Coxaldehid]   and the changing of that

Table 10-5   *Effect of different methods of saffron drying on the level of smell/ aroma composition (38)*

S.No.	Drying method	Volatile oil (mg/g)	Saffranal in volatile oil (mg/g)	4-Beta Hydroxy saffranal in volatile oil (mg/g)
1.	Fresh saffron (wet)	Partly	-	-
2.	Drying in shade, sun or by electric oven (at 65°C, 50°C, 40°C & 20°C)	6.0	550-680	140-200
3.	Drying by blowing air in the opposite direction or under vacuum	5.0	200-350	500-700

to glucose, saffranal *(2 & 6 & 6- Tri-methyl-1-3- Cyclohexa D- N -1- Carboxaldehid)* and 4-Hydroxy saffranal. Resulting safranal changes to izaforon by enzymatic and non–enzymatic reactions, decarboxidation and isomerisation during storage.

Hemmati Kakhaki (13) has also examined, in a series of experiments, the effect of all four abovementioned methods of drying saffron including the traditional Iranian method, the Spanish process (using of the sieve of silken threads), electric oven and also drying in vacuum on the main compositions of saffron (the compositions, which influence the color, taste and smell of saffron). In the light of these results he has proved that the color of saffron can be maintained better by drying it in an electric oven at 60°C for a period of 2.5 hours compared with other methods. The Spanish system stands next to this system as far as maintaining the peculiarity of the quality of saffron is concerned (Fig. 10-2). The results of this research show that one of the main factors that cause the degradation in the quality of saffron is the drying process. Long duration of drying saffron gives an opportunity to enzymes to act and analyze the color stuff. Besides, favorable conditions for the growth and development of microorganisms (moisture, heat and substance), their number multiplies enormously. After these experiments, it has

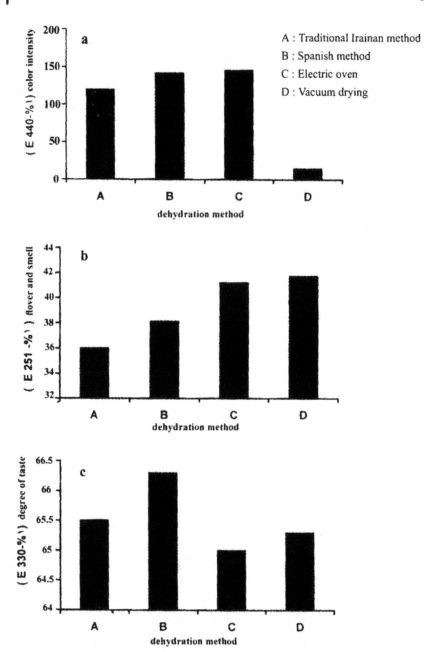

Fig. 10-2 The effect of different methods of drying in preserving the qualitative peculiarities of saffron- a) Color; b) Smell and flavor; c) Taste (-13)

been recommended that in the scattered regions and small farms the Spanish method (being simple and less expensive) is quite suitable. While in bigger farms and the regions where this industry is concentrated, the industrial drying system is suitable with controlled circulation of air for drying saffron.

In Italy, after separating the stigma from the point where it is joins the style, saffron is carried in sieves made of silken threads. It is placed on a stand at a distance of 20-cm from the source of heat, which is wooden charcoal, so that it dries. This process takes about 15 to 20 minutes. It is believed that the saffron dried by this process has bright crimson color as compared to electric oven dried saffron and has a better taste. From each 500 g fresh stigma, 100 gm dry saffron is obtained with 2–5% moisture. The dried stigmas are put in a coffee making electric grinding mill to extract the powder. The powder, which is produced by this method is kept in covered glass jars and is placed at a dry place (32).

In Greece, for drying saffron, first the fresh saffron is placed in layers of 4-5 mm on plates of 50×40 cm diameter, covered with silken cloth. After that these plates are set over each other in a firm cupboard at a distance of 25–30 cm and are placed on a stand. These cupboards are kept in a dark room where there is enough heat to dry the product. The temperature of the room can be regulated. At early stages of drying, the temperature is kept at 20°C, afterwards it is raised to 30°C and then 35°C. When the moisture content of the product is reduced to 10-11%, the heating is stopped.

The time taken for drying saffron by this method is 12 hours. After cooling, the saffron is sorted out and all the extra matter is separated. Now, the pure dried saffron is kept in uninfiliterated glass bottles or metal caskets and the temperature maintained at about 5-10°C.

In Morocco, saffron is dried immediately after separating it from the flower. It is done by spreading the stigmas in thin layers on a sheet of cloth and kept under direct sunlight or in the shade. It takes two hours to dry saffron under direct sunlight whereas in the shade it takes 7-10 hours (32).

On the basis of the recommendations made by the National Standard Organization of Iran No. (22) 5230, the flowering points should be observed, so that the product be hygienically fit and with suitable quality.

- Indirect heat should be used to dry saffron
- That process of drying should be chosen which takes shorter time to dry the saffron
- At the time of drying, the temperature should be uniformly maintained and it should not go beyond 60°C
- For drying purposes metal surfaces and unhygienic plates should not be used
- The final moisture content of saffron, i.e. when the drying process is over, should not be more than 10%.

10-2-3 Packing and Storage of Saffron

When saffron is dried and sorted, it is ready for packing and display. The major points that should be kept in mind while storing saffron are listed below, which otherwise may spoil or degrade in quality:

- The moisture of the product and the relative humidity of the store
- The temperature of the place where saffron is kept
- Light, particularly direct sunlight
- Oxygen
- Type of container.

Rahimi (26) has tested the effect of temperature and moisture of saffron to maintain its physico-chemical and microbic properties. In this research saffron was kept at three different moisture levels (6%, 10% and 14%) for one year at varied temperatures (4, 20 and 40°C). During this period, the evaluation of samples showed that the conditions of maintaining saffron have great effect on the balance of color and smell combinations. If the temperature of a place where saffron is kept is low and the moisture of the product is less, the quality of saffron is better. This examination showed that when the saffron is well maintained, the number of microorganisms reduces gradually. However, the process of decreasing in samples, which were kept at the temperatures of 20°C and 40°C is stronger than that of the temperatures of a refrigerator (6 ± 2).

Noor Baksh and his associates (33) examined the packing and temperature to maintain physico-chemical and microbic properties of saffron. The results of these tests showed that method of packing may have a meaningful effect to maintain the physico-chemical properties of saffron (color, taste and aroma). Four kinds of packing were tested, for which glass bottles, low-density polythene, high-density polythene and layered aluminum foils were used. It was found that glass bottles and polythene (of any density) have better preserved the qualitative properties of saffron, compared to layered foils. But the different kinds of packing had no noticeable effect on saffron pollution.

The results of Hemmati and his associates (12) on the kind of packing to maintain the quality of saffron showed that the samples of saffron that were kept in the colored glass bottles had more color strength than the samples kept in cardboard containers, high and low-density polythene packets.

Raina and his associates (38) reported that saffron at a moisture of about 5% is maintained longer in the polytyrene caskets coated with selephone MXXT. Monnio and Ameioth (28) studied the effect of temperature and moisture of the store on maintaining the qualitative properties of saffron. They reported that the stability of color in low relative humidity (5-23%) remains better, but the stability of the color at 75% relative humidity is very little. When saffron is maintained at low temperatures the durability of color is longer. Morimoto and his associates (30) reported that during saffron storage depending on level of heat, humidity and oxygen glycozyl esters of crocine are analyzed and quality of saffron reduces consequently.

10-2-4   Removing of Pollutants from Saffron

Almost all the spices including saffron are strongly polluted by microorganisms. That is why the consumers who use these products, particularly those who are in the food industry and use the spices on a regular basis, are facing difficulties. As the saffron flower grows on the surface of the soil and is harvested from there, it leaves ample opportunity dust and fertilizers to pollute the flowers. When the technical and hygiene principles are not observed during cultivation, harvesting, transportation, preservation of flowers, separation of stigma and drying of saffron, there is intensive increase in microbes.

Microbic flora, which exists in almost all the spices, contains the aerobic spleen producing bacteria and moulds. Sometimes strapetcocks and chliforms are found in it, however, clostridiums, lactobacillus, microcucks, staphilocuks and yeast are hardly found.

The special peculiarities of saffron are its color, taste and smell. These are the properties that give it particular worth and value. But removing pollutants, high temperature decreases quality and that is the reason that among all the different methods to maintain the quality, the more appropriate process is the cool methods. For this purpose, using ion making rays, microwaves, ultra-violet waves and fumigation can be used.

The usage of irradiation in the food-industry is comparatively a new method. Due to the changes that take place in the atoms of radioactive elements, alpha, beta and gamma rays are produced. These rays have wave-length less than 1000 Å, and the capability of killing microbes. The penetration rate of the above-mentioned rays is also different. The particles of alpha rays cannot pass through even a piece of paper. The particles of beta rays with electrons have more capability of penetration, but they are not able to pass through aluminum sheets. The gamma rays have a very high capacity of penetration. In the process of applying rays, food articles need those rays that are more able to penetrate. These do not cause to analyze the compositions of the food–stuff and also make them radioactive. It is for this purpose that gamma rays are being used. The gamma rays are taken from radioactive elements like Uranium. The food article is kept for a fixed time under the rays with particular circulation so that the microorganisms are removed.

Contrary to ionized waves, microwaves are used for heating food. In this process, the phenomenon of the cooling point does not remain the same in the foodstuff. This is because all the points get heated without difference of heat. The ultra-violet rays, at the wavelength of 2000-2800 Å have the highest capacity of killing the microbes. But their penetration capability is quite meager.

In the process of fumigation, by using gases, fumes and different vapors, insects and microorganisms are removed from the food articles. The fumigant of food, in order to sterilize it, must have some peculiar properties. Among them is the broad spectrum, which takes care of most of the microorganisms and its application is easy and also meets the minimum requirement to remain in the foodstuff. Tests and studies have revealed that Ethylene-oxide is an effective compound for fumigation.

Hemmati and his associates (12) have tested the effect of the two methods of fumigation and gamma rays to reduce the microbic content in saffron. In the first process, ethylene oxide was used at three different densities (250, 500, 750 ppm) for twelve hours to reduce the microbic level in saffron. The results showed that different concentrations of ethylene oxide had no undesirable effect upon the peculiarities of color, taste and smell of saffron. The density of fumigation further reduced the level of pollutants.

In the second system of removing pollutants from saffron, gamma rays have been applied at three levels of 2000, 4000 and 6000 gray. The results showed that, although the gamma rays at 6000 gray cause reduction in microbic quantity but the process also decreases the color strength of saffron.

10-3   CHEMICAL COMPOSITION OF SAFFRON

10-3-1   General

The type and amount of saffron components is shown in Table 10-6 (32). Saffron carbohydrates are mostly from reductive sugars, which consist of around 20% of saffron dry weight. Amongst these carbohydrates the presence of glucose, fructose, gentibiose and small quantity of xylose and ramnose were fixed. Table 10-7 shows all types of saffron carbohydrates and pigments.

Table 10-6   *Analytical results of main components of commercial saffron (10)*

Type of Component	Size (w/w %)
Moisture	10
Carbohydrates (based on invert sugar)	14
Tannins	10
Pentosanes	8
Pectin	6
Starch	6
a-Crocine	2
Other Carotenoids	1
Proteins (N X 6.25)	12
Mineral elements (ash)	6
Non-soluble in acid ash	0.5
Non-volatile oils	6
Volatile oils	1
Crude fibers	5

Total ash of saffron varies from 4% to 8%. Kuntze and Higler (in 10) analyzed mineral elements of saffron and found the percentages given below for each compound:

$$Cl = 2.88\%$$
$$SO_3 = 7.12\%$$
$$P_2O_5 = 10.01\%$$
$$Na_2O = 8.56\%$$
$$K_2O = 34.46\%$$

Table 10-7 *Different kinds of carbohydrates and pigments of dry saffron stigma (10)*

Type	Sample-1 (%)	Sample-2 (%)
Total reductive sugars	26.70	26.30
Free reductive sugars	14.30	13.80
Sugars after inversion	14.80	14.30
Tannins & dexterine	9.40	10.30
Starch	6.00	6.40
Pentosane	6.40	6.90
Free crocine	0.60	0.50
Crocetine	8.40	8.80
Non-measurable materials	6.70	6.40
Gentibioside	2.79	2.31
Glucose	7.59	7.88
Fructose	1.82	1.91

International Standard Organization reported that the total weight of different elements in 100 g of dried saffron as: calcium 111 mg, ph osphorous 525 mg, potassium 1724 mg, sodium 148 mg, zinc and magnesium in small quantities (18).

The total ash and acid, insoluble ash of saffron is measured as one of the main factors of quality as per international and national standards. The ash content of saffron is also important for its purity (11).

Saffron is one of the richest sources of riboflavin. The amount of thiamine and riboflavin in four samples of saffron is shown in Table 10-8. In one study white mouse was treated with 150 mg of saffron and the effect was similar to treating the same mouse with 40 mg of synthetic riboflavin.

10-3-2   Specific Composition of Saffron

One of the main purposes of saffron consumption is its color property (4). Pigments of saffron are mainly from carotenoids group with carboxyl. These pigments are present as free crocetin 8, 8^1-diaparorotene- 8-8^1-dicarboxilic acid and other forms along with

Table 10-8   *Riboflavin and thiamine of four samples of saffron*

Number of sample	Riboflavin ($\mu g/g$)	Thiamine ($\mu g/g$)
1	138.0	4.0
2	93.0	3.8
3	78.70	0.72
4	56.40	0.88

glucose gentibiose as glycosyl esters. The main coloring property of saffron comes from crocetin. Many investigations revealed that 94% of total crocetin of saffron is present in the form of glycosides compounds in crocetin and the remaining 6% is in the form of free crocetin (Fig 10-3).

The coloring power of saffron comes from the above-mentioned compounds, which are measurable by different methods, for instance, spectrophotometry, high performance liquid chromathography (HPLC) and thin layer chromathography (TLC).

As per international standards as well as many national standards specifications, coloring strength of saffron is expressed as direct reading of the absorbance of crocine at about 440-mm on dry basis (17, 21).

It seems that aging of saffron, particularly under unfavorable conditions, gradually causes increasing of mono-esters and free crocetin and subsequently decreasing of di-ester crocetin, which subsequently reduces the color strength. (24).

Orfanou and Tsimdou (36) evaluated the effects of different solvents on extracted color from saffron stigmas and reported that a mixture of water and ethanol in equal proportion, extracted the maximum color from saffron stigmas. Table 10-9 represents the results of this experiment.

Pfander and Rychaner (37) applied different methods for separation and extraction of glycoside ester crocetins from ethanol extract of saffron using HPLC. They reported that after filtration of crude extract of sephdex G-50 using lichrosorbsi 60 as fixed phase and ethyl acetate, iso propanol and water (63:34:10) as movable

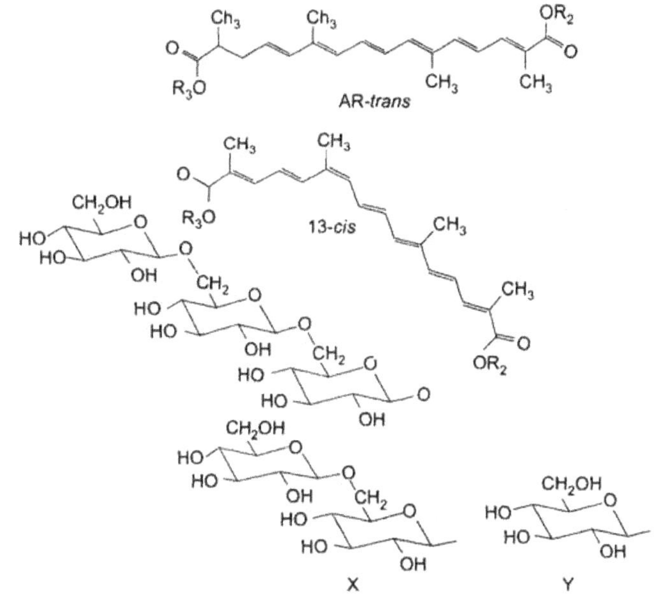

Fig. 10-3 Chemical structure of saffron stigma

Crosins (CRCS) Glucosylesters of crocetin
Crosin-5: R1 = three B-D-glucosyl (z) R2 = B-D gentrobiosyl(x) $C_{50}H_{74}O_{29}$ MW : 1138
Crosin-4: R1 = R2 = B-D-gentiobiosyl(x) $C_{44}H_{64}O_{24}$ MW : 976
Crosin-3: R1 = B-D-gentiobiosyl (x) R2 = B-D-glucosyl (Y) $C_{38}H_{54}O_{19}$ MW : 814
Crosin-2: R1 = B-D-gentiobiosyl (x) R2 = H $C_{32}H_{44}O_{14}$ MW : 652
Crosin-2: R1 = R2 = B-D-glucosyl (Y) $C_{32}H_{44}O_{14}$ MW : 652
Crosin-1: R1 = B-D-gucosyl (Y) R2 = H $C_{26}H_{34}O_9$ MW : 490
Crocetin (CRT): R1 = R2 = H $C_{20}H_{24}O_4$ MW : 328

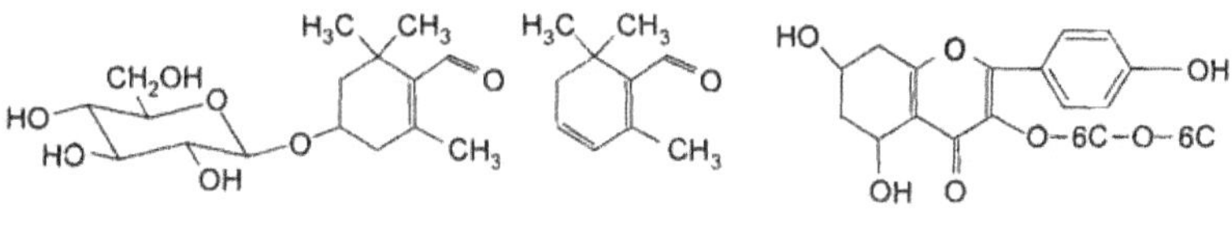

Picrocrocin
$C_{16}H_{36}O_7$ MW : 330

Safranal
$C_{10}H_{22}O$ MW : 150

Di-glucosyl-Kaempferol
$C_{27}H_{30}O_{16}$ MW : 610

phase produced the best results for separation of coloring pigments of saffron.

Himeno and Konosuke (15) first extracted these compounds in ethanol 50%. In order to measure the chemical composition of saffron using HPLC and then determining different compounds

Table 10-9 *Effects of type of solvent on the concentration of colors extracted from saffron stigmas (36). Each number is the average of five samples*

Solvent	Color strength $E^{1\%}$ 442
Cold water	212.6 ± 0.9
Water + Ethanol (1:1)	222 ± 0.9
Water + Ethanol (1:4)	213.5 ± 1.9
Water + Methanol (1:1)	228 ± 4

using reverse phase column and UV visible detector they found 7.2% crocin and 1.62% picrocrocin in fresh saffron but no measurable safranal.

Tantillis et al. (42) determined saffron composition using HPLC and mass spectrometry of which the results are shown in Figure 10-4. Picks are: 1- picrocrocin; 2. acid from picrocrocin; 3. campferal diglycoside; 4. trans-crocin; 5. trans-crocin-4; 6. trans-crocin-3; 7. safranal-1; 8. trans-crocin-2; 9. cis-crocin-5; 10. cis-crocin-4; 11. trans-crocin-2; 12. cis-crocin-3; 13. cis-crocin-1. Ibora et al. (16) introduced a method for extraction and separation of saffron compounds. They first evaluated the effectiveness of solvents such as ethyl ether, acetone, acetonitryle, methanol, ethanol, isopropanol, ethanol 50% and water and concluded that water and ethanol 50% produce that best results (Table 10-10).

They separated different compounds of saffron using thin layer chromatography on aluminum oxide sheets and reading absorbance by spectrophotometer. The quantitative composition of saffron pigments was determined using HPLC and reverse column C-18 and under UV visible detector. The results were shown in Table 10-11 (16).

Factors like temperature, oxygen, light and acidity can affect pigment stability of saffron. Tsimidou and Tsatsaroni (43) studied the effects of the above-mentioned factors for which the results are as follows:

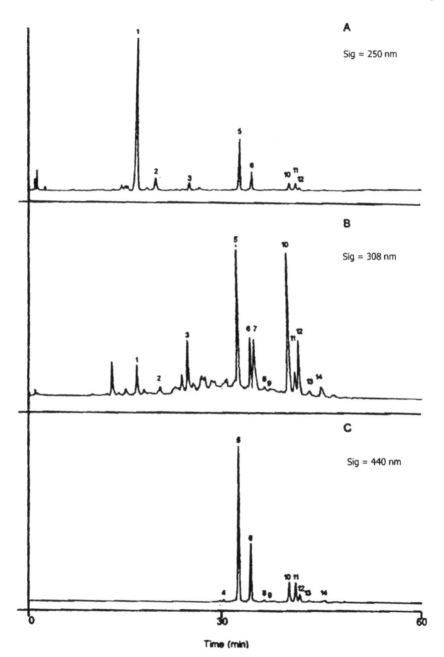

Fig. 10-4 Separated saffron compounds in Methanol 5% extracted by HPLC in three wave lengths (A) 250 nm, (B) 380 nm and (C) 440 nm

Table 10-10 *The extracted amount of picrocrocin, crocin and HTCC from 10 mg of saffron by water and Ethanol 50% as solvent. The numbers in parenthesis are the percentage of the compound based on dry extract of saffron*

Solvent	Picrocrocin µg	HTCC µg	Crocin µg
Water	1385 (13.9)	141 (1.4)	799 (8)
Ethanol 50%	1293 (12.9)	132 (1.3)	786 (7.9)

Table 10-11 *Analysis of saffron by HPLC, spectrophotometer and thin layer chromatography (TLC) (16)*

Compound	TLC R$_f$	HPLC Time prevention (Min.)	Spectrophotometry λ_{max} in water (nm)
HTCC	0.86	6.8	250.5
Picrocrocin	0.76	4.0	250.0
Crocin ester-1	0.60	6.7	440.0
Crocin ester-2	0.33	5.7	440.0
Crocin	0.13	5.0	440.0

PH and Light effects:

The effects of H^+ on stability of saffron pigments in aqueous extracts were studied at pH 3, 5 and 7 at 40°C, and the results shown in table 12-10. Based on these results stability of saffron carotenoids was pH dependant and pigment degradation is higher in pH=3 compared with pH =7. The combined effects of pH and light on degradation of pigments showed there was no significant difference between light and dark on pigment degradation (Table 10-13). Reduced atmospheric oxygen also presented a minimal protection on crocin shelf life under all conditions. This result may be related to lower rate of oxidation of carotenoids with relatively higher molecular polarity.

Thermal stability of crocin was studied at four temperatures (4°C, 25°C, 40°C and 62°C) at pH =7 in the dark in the presence of oxygen (Table 14-10). Incubation at 4°C reduced the degradation rate by factors of more than 3, 8 and 10 compared to reaction rate constants at 25°C, 40°C and 62°C respectively.

Table 10-12 *Reaction rate constant and calculated half-life periods of degradation of saffron pigments aqueous extract at 40°C.*

Experimental conditions	Measurements	Correlation coefficient*	$K \times 10^{-3} \pm S_K \times 10^{-3}$ (hr^{-1})	Half-life (hr)
Air pH=3	6	–0.984	43 ± 6.3	16
Air pH=5	10	–0.992	7.0 ± 0.7	99
Air pH=7	12	–0.994	5.8 ± 0.2	120
Nitrogen pH =3	6	–0.961	39.7 ± 16.5	18
Nitrogen pH= 5	10	–0.996	6.8 ± 0.7	102
Nitrogen pH =7	12	–0.989	5.0 ± 0.5	139

Ln E $^{1\%}$ $_{440}$ = Const. + K+

Table 10-13 *Combined effects of pH and light on degradation of saffron pigments aqueous extract at 40°c, first-order reaction rate constants and calculated half-life periods*

Experimental Conditions	Measurements	Correlation coefficient*	$K \times 10^{-3} \pm S_K \times 10^{-3}$ (hr^{-1})	Half-life (hr)
Light, Air pH = 5	10	-0.987	8.0 ± 1.1	87
Dark, Air pH = 5	9	-0.995	11.7 ± 0.7	59
Light, Air pH = 7	8	-0.987	5.9 ± 0.6	117
Dark, Air pH = 7	7	-0.997	8.4 ± 2.0	84
Light, N_2 pH = 5	9	-0.982	11.0 ± 4.5	63
Dark, N_2 pH = 5	9	-0.990	11.3 ± 0.3	61
Light, N_2 pH = 7	7	0.945	8.2 ± 2.0	85
Dark, N_2 pH = 7	7	0.982	7.7 ± 1.4	90

* Ln E $^{1\%}$ $_{440}$= Const. +K+

Table 10-14 *First order reaction rate constants and calculated half-time periods of saffron pigments aqueous extract at 40°C.*

Experimental Conditions	Measurements	Correlation coefficient*	$K \times 10^{-3} \pm S_K \times 10^{-3}$ (hr^{-1})	Half-life (hr)
Air pH =3	9	0.996	4.0 ± 0.7	175
Air pH= 5	12	0.986	0.9 ± 0.4	779
Air pH= 7	10	0.968	0.4 ± 0.2	1873
Nitrogen pH =3	9	0.993	3.8 ± 0.2	180
Nitrogen pH= 5	12	0.973	0.9 ± 0.2	770
Nitrogen pH= 7	10	0.939	0.3 ± 0.06	2235

*Ln E $^{1\%}$ $_{440}$= Const. +K+

10-2-2-3 *Factors responsible for flavor, aroma and smell*

Saffron tastes bitter and is slightly pungent. The main compound that gives taste to saffron is picrocrocin, which is a colorless glycoside present around 4% in fresh stigma. Figure 10-5 shows the space structure of picrocrocin (10).

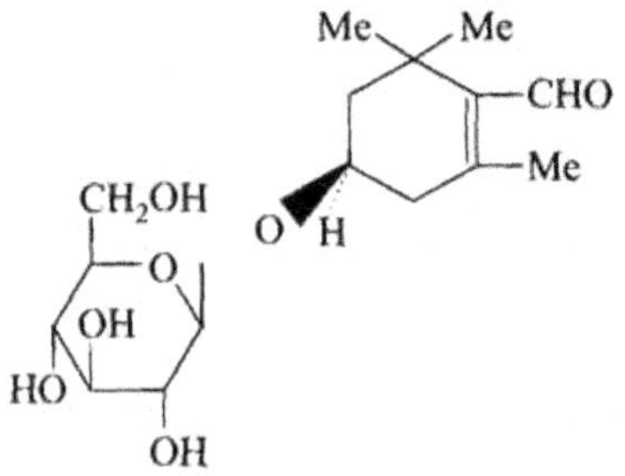

Fig. 10-5 Space structure of picrocrocin (10). After the analysis of picrocrocin, safranal, a volatile aglicone is produced.

$$C_{16}H_{26}O_7 + Ba(OH)_2 \rightarrow C_{10}H_{14}O + C_6H_{12}O_6$$

Picrocrocin   Steam   Safranal   Glucose

Himeno and Konosvke (15) in their experiment first separated picrocrocin from ethanol extract using HPLC (Fig. 10-6) and then one part of picrocrocin is mixed with normal HCl and other part is mixed with normal NaOH. The mixture is kept at 37°C, picrocrocin is then analyzed in this condition and changes to safranal. They also used beta-glycosidase for analyzing picrocrocin. The results are shown in Fig. 10-7.

The aroma and smell of saffron is because of volatile oils. By distillation of saffron by distilled water along with carbon dioxide and mixing the result in ether and distillation of ether under carbon dioxide, these oils will be separated. The separated oil is transparent and belongs to terpen group that absorbs oxygen easily and changes to a dense brown liquid.

Saffron oil with density of 0.9514 to 0.9998 is a little bit left inclined and without shaking it will precipitate and produce

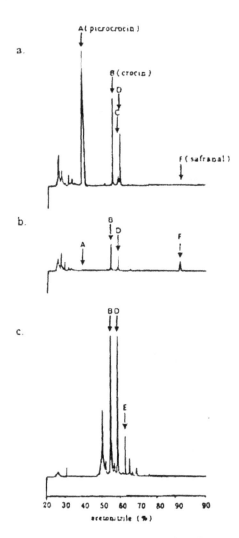

Fig. 10-6 Chromatograms of ethanol extracts of saffron, separated by HPLC a:250.5 nm, b: 380 nm and c: 443 nm (15). Picks are A: Picrocrocin, B: Crocin, C: HTCC, D: Unknown and F: Safranal

stearoptene crystals, which after re-crystallization, these crystals in ether de-petrol will show a melting point of 106°C (10).

Safranal is the main aromatic component of volatile oils of saffron, which is produced due to hydrolysis and rehydration of picrocrocin (15). Zarghami and Heinz (47) extracted the volatile

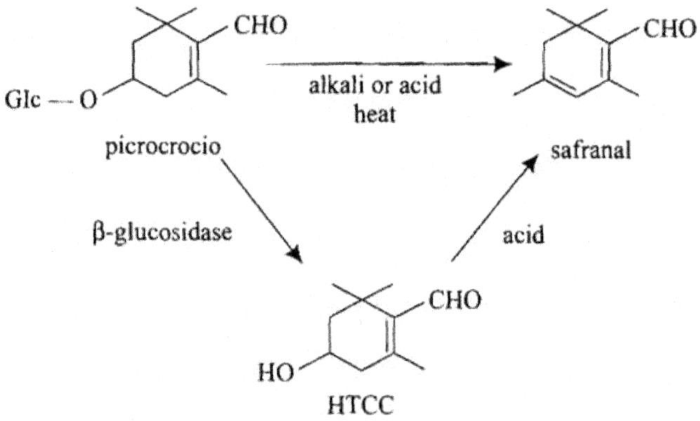

Fig. 10-7 Change of picrorocin to safranal and HTCC by chemical and enzymatic methods (8)

compounds of saffron by using diethyl ether and separation and recognition of saffron essential oil components was done by gas chromatography, far-red spectroscopy, ultra violet and nuclear magnetic resonance. For the first time they reported seven other compounds except safranal in saffron essential oil. Name and chemical structure and relative density of these compounds are given in Table 10-15.

There are some differences in safranal content of saffron essential oils. Zarghami and Heinz (47) reported 47% while Rodel and Petrzita (6) reported 60.3% and Alonso et al. (25) 72% of safranal content in saffron oil. Some reports indicate that safranal has got anti-bacterial trait. Noorbaksh et al. (33) studied *E. coli*, *Staphylococous Aaurus*, *Psoudomunal Aerojeneusa* growth in agar using three-dimensional microbial *E. coli* and applied saffron water extract containing crocin, crocetins, picrocrocin and safranal and found that water extracted saffron had no preventive effect on micro-organisims, however, safranal showed negative effect on growth of *E. coli* and staphylocoke. Alonso et al. (2) suggested that for measuring aromatic compounds of saffron particularly safranal HPLC and spectrophotometry are not precise methods. They found that gas chromatography is more precise.

Table 10-15 *Chemical structure and relative density of volatile compounds of saffron compared to safranal (47)*

Compound	Structure	Relative Density based on Safranal = 100
1	Iso-Phorone	3.94
2	3,5,5 Three Methyl-4-Hydroxi 1-Cyclohexayoun-di n	12.8
3	3,5,5 Three Methyl-4- Cyclohexayoun-di n	3.28
4	3,5,5 Three Methyl-4- Cyclohexayoun-di 2n	2.47
5	3,5,5 Three Methyl-2- Hydroxi 1,4 Cyclo, de n-2-n	2.47
6	2,6,6, Three Methyl-4-Hydroxy-2-Cyclo Hexen-1-Carboxaldehi	29.4
7	2,4,4 Three Methyl-3-Formil-6-Hydroxi, 2,5, Cyclohexa-DN-1-N	12.8

Quality of saffron in international trade is determined based on International Standard Organization ISO 3632 document (17). This standard measures chemical and microbial characteristics of saffron and based on optical density of water extract, the quality level of saffron is determined. Saffron standard suggests that optical density of water extract should be obtained at 440 nm for crocins (pigments), at 257 nm for picrocrocin (flavor compound), and at 330 nm for safranal (aroma and smell), but this method is still under debate (Fig. 10-8).

Trantillis et al. (42) showed that crocin also has absorbance in 257 and 330 nm. In addition, cis crocin isomers also have absorbance in 330 nm. Since saffron compounds are not polar molecules, therefore, they are not completely solvable in water and ISO standard for saffron should be revised.

10-3 CHANGES IN SECOND METABOLITES DURING FLOWER EVOLUTION IN SAFFRON STIGMA

Table 10-16 shows changes in crocin, picrocrocin, safranal and HTCC in saffron stigma at flower completion. In the first stage, 7

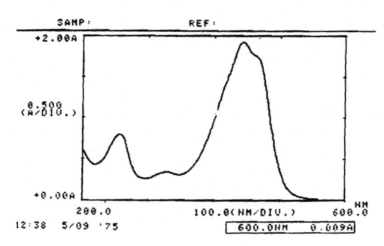

Fig. 10-8 Absorbtion bound of saffron water extract

Table 10-16 *Changes in second metabolites in saffron stigma during flower evolution (8)*

Stage of stigma evolution	Stigma Length (cm)	Stigma Color	Flower Age (weeks)	Stigma fresh weight (mg)	Crocin (%)[1]	Picrocrocin (%)	Safranal (%)	HTCC (%)
1	0.1	Color-less	-7	0.03	0	0	0	0
2	0.2	Light yellow	-6	0.2	0.05	0.01	0	0
3	0.3	Yellow	-5	1.1	0.91	0.16	0	0
4	0.5	Orange	-4	3.1	2.37	0.44	0	0
5	0.8	Red	-3	3.5	7.20	1.43	0	0
6	0.1.2	Red	-1.5	5.3	7.17	1.62	0	0
7	2.2	Red	-0.5	8.8	5.64	1.14	0	0
8	2.2	Red	0	13.0	3.32	0.25	0.00038	0.0046

[1]Percent is based on fresh weight of stigma.

weeks before flowering, the stigma is completely colorless and none of second metabolites are mesurable. In the second stage, 6 weeks before flowering, the stigma becomes a little yellowish and small amount of crocin and picrocrocin is detectable. In stages 3 to 5, with changing of stigma color from yellow to red, the content of crocin and picrocrocin increases rapidly. From stages 5 to 7 the rate of crocin and picrocrocin increment is slower than the previous stages.

The ratio of crocin to picrocrocin in all stages of flower evolution is almost fixed. In all seven stages safranal and HTCC is not detectable. In stage 7 (flowering) considerable safranal and HTCC is accumulated. It is exceptional for stigma to accumulate these second metabolites and they are not accumulated in leaves, roots, corms, petals and stamens (8).

10-4 STANDARDS AND QUALITY CONTROL OF SAFFRON

10-4-1 Saffron Standards in Iran

10-4-1-1 Saffron characteristics

Saffron is a very expensive spice and for a few grams customers have to shell out quite hefty sums (9). It is, therefore, important that the quality of the product matches national and international standards. At international level, International Standard Organization is responsible for ensuring standards for each product and at the national level in Iran, Institute of Standard and Industrial Research Organization (ISIROI) is responsible for guarding the quality standards of products including saffron.

The main aim of saffron standard is to specify the norms for packing, labeling and sampling and for methods of testing and whether the particular produce is fit for filament or powder form. In this standard after definition of terms, specification of standard saffron was introduced and published under the title 'Saffron Specifications' under registration number 259-1 (20).

There is another part of the standard, which introduces the methods suitable for testing saffron and it is applicable to filament and saffron powder. In this issue, besides explaining general tests such as moisture and volatile matter, color strength, bitterness, flavor, floral waste content, microscopic examination of saffron powder, determination of total ash and acid-insoluble ash, measurement of crocin, picrocrocin and safranal by spectrophotometry is also included. This publication in the name of 'Saffron Test Methods" under the number 259-2 was released by Standard and Industrial Research Organization of Iran (SIROI). There is one more standard in the name of microbial specification of saffron, which is published by ISIROI under the title 'Microbial Specification and Its Tests' under the number 5689 (23). This standard examines any microbial contamination during picking flowers, flower transportation, stigma separation, drying and packing. Based on this standard the maximum population of microorganisms in saffron must not exceed the numbers given in Table 10-17.

Table 10-17 *Microbial specification of saffron (23)*

Microorganisms	Maximum number per gram	Method of testing
Total microorganisms	5×10^4	Based on number 5272-SIROI
Coliform	10^3	Based on number 437-SIROI
E. coli	0	Based on number 2946-SIROI
Chlostridium per	0	Based on number 2197-SIROI
Mould	10^3	Based on number 997-SIROI

ISIROI released a publication 'Control Points of Harvesting and Processing of Saffron' under the number 5230. The aim of this control point was to introduce Suitable methods for harvesting, separation, drying and storing of saffron. This book suggests the best time for flower harvesting, methods of harvesting, the flower collectors, transportation and flower container.

The Hazard Analysis Critical Control Points (HACCP) is also implemented by some companies dealing with saffron particularly drying, packing and exporting saffron. In this standard after definition of terms, it states the best way of establishing a system for analyzing hazardous risks and control critical points from the time of flower harvesting to the time of stigma packing. Figure 10-9 shows the procedures and critical control points and control of these points (24)

Protocol for starting up packaging plants of saffron is also standardized by ISIROI and it is compulsory for companies that wish to establish packaging plants. The purpose of this protocol is to respect the main principles, which are important to introduce a hygienic product in the market. This book has two parts. The first part explains the characteristics of location where the packaging plant could be set up, structure and plan for different assemblies in the plant, quality control laboratory, stores, hygiene facility, and health certificate of workers. The second part of the book describes the methods and different phases of packaging (Fig. 10-10). The last standard of saffron is the method of sampling, which is published in the number 3659 by ISIROI. In this standard after the definition of terms, describes sampling from packed saffron.

10-4-2   International Standard of Saffron

Saffron being a very expensive spice must fulfill at least the prescribed standard of the country of origin for export to international markets. International Organization for Standardization (ISO) is a worldwide federation of national standard bodies for saffron. ISO3632-1 and 3632-2 were prescribed by a technical committee: Agricultural food products subcommittee, spices and condiments. The ISO 3662 consists of specification and test methods parts.

Part one of ISO 3632 specifies the requirements for saffron obtained from the flowers of *Crucus sativus* Linnaeus. It is applicable to whole filaments and in powder form. In this standard filament saffron is classified into four groups (Table 10-18) based on floral waste and extraneous matters.

Chemical characteristics of saffron based on international standard. ISO represented in Table 10-19 (17). The methods

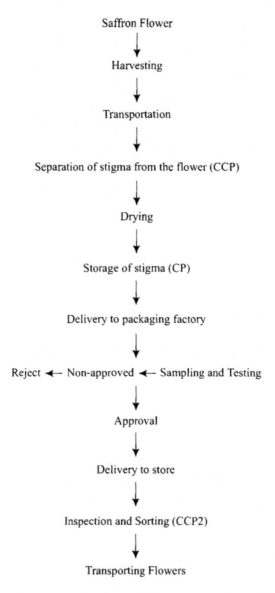

Fig. 10-9   Flow chart of control points and critical control points from harvesting to packaging of saffron. CP=Control Points, CCP=Critical Control Points (24)

presented in this standard were used preparing Iranian Standard Number 259-2.

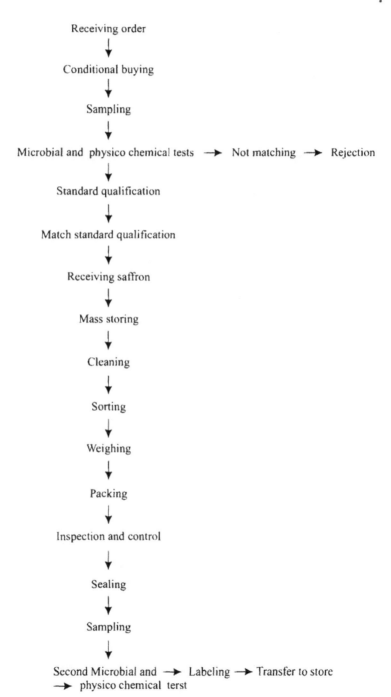

Fig. 10-10 Different phases for saffron packaging (25).

Table 10-18 *Classification of saffron in filament*

Characteristics	Categories			
	I	*II*	*III*	*IV*
Floral Waste % (m/m) Max.	0.5	4	7	10
Extraneous matter % (m/m) Max.	0.1	0.5	1.0	1.0

10-4-3   Saffron Standards in other Countries

There are different scales and standards in many countries, which are important for saffron producers and exporters. Here we mention some important standards in different parts of the world.

10-4-3-1   *Scales and standards of saffron in USA*

Based on FDA standards, the minimum specification of saffron in USA should be as under:

1. Dried yellow stamens and extraneous matter of saffron should not be more than 10%
2. When saffron is dried in 100°C, its volatile and moisture content should not exceed 14%
3. The total ash content of saffron should be less than 7%
4. The acid insoluble ash should not be more than 1%

Based on these standards saffron could be used as natural coloring agent and there is no limitation to use it in food material.

10-4-3-2   *Prevention of food adulteration act of India*

Based on this Act saffron should be free of any external coloring agents and should have specifications mentioned below:

1. Total ash content is not more than 8% of the total weight
2. Acid insoluble ash should not be more the 1.5% by weight
3. Moisture and volatile matter (103 ± 1°C) should be less than 14%
4. Aqueous extract should not be less than 55%

Table 10-19 *Chemical characteristics of dry filament and powder of saffron based on ISO 3632-1 (17)*

Characteristics	Requirements		Test Methods
	Filament	Powder Form	
Moisture & volatile matter % (m/m) max.	12	10	ISO 3632-1 Clause 9
Total ash % (m/m) on dry basis max.	8	8	ISO 928 & ISO 3632-2 Clause 10
Acid insoluble ash % (m/m) on dry basis max.			ISO 930 & ISO 3632-2 clause 11
Categories I & II	1.0	1.0	
Categories III & IV	1.5	1.5	
Solubility in cold water % (m/m) on dry basis max.	65	65	ISO 941
Bitterness expressed as direct reading of the absorbance of picrocrocin at about 257 nm, on dry basis min.			ISO 3632-2 clause 13
Categories I	70	70	
Categories II	55	55	
Categories III	40	40	
Categories IV	30	30	
Safranal expressed as direct reading of the absorbance at about 330 nm. on dry basis			ISO 3632-2 clause 13
All categories Min.	20	20	
Max.	50	50	

(Table 10-19 Contd.)

(*Table 10-19 Contd.*)

Coloring strength expressed as direct reading of the absorbance of crocin at about 440 nm on dry basis, Min.			
Categories I	190	190	
Categories II	150	150	
Categories III	110	110	
Categories IV	80	80	
Total nitrogen % (m/m) on dry basis Max[1]	3.0	3.0	ISO 1871
Crude fiber % (m/m) on dry basis Max[1]	6.0	6.0	ISO 5498

[1]Additional tests, which may be carried out if necessary. If sufficient sample is available.

5. Total nitrogen (based on dry weight) should be less than 2-3%

6. Foreign materials such as soil, sand, leaves, stem and straw should be less than 1% by weight

7. The flower components such as stamens, pollens and ovary should not be more than 15% in saffron

10-4-3-3 Saffron standard in the United Kingdom

British Standard Institute (BSI) accepted ISO 3632 as British saffron standard (6, 7)

10-4-3-4 Methods of standardization and test of saffron in Spain

In this document all methods of quality measurements and identifying pure and low quality saffron is described. Based on this standard saffron is classified into six grades as are given below (39, 41):

1. **Mancha saffron:** The length of stigma must be more than the length of style and the color strength of this grade must be more than 180 units. The maximum flower content of mancha should not exceed 4% by weight

2. **Rio saffron:** The length of stigma should be at least equal to style, which is connected to it, the color strength should be more than 150 units and the extraneous flower component (except stigma and style) should not be more than 7% by weight

3. **Sierra saffron:** Length of stigma is less than style, maximum 10% of flower component and coloring strength should not be less than 130 units

4. **Common saffron or Standard:** This saffron is a mix of above three types, color strength should not be less than 130 units and the remaining flower parts should be less than 7% by weight

5. **Cupe saffron:** This saffron is free of style and it is pure stigma, minimum coloring strength of this type of saffron should be 190 units

6. **Molido saffron:** It is the powder of above saffrons and it should be mentioned which type of saffron and its specifications match with this type of saffron. The moisture content of this saffron must be less than 8%

10-4-3-5 Saffron Standard in France

The French Standard Organization (AFNOR) released saffron standards in 2000. This standard consists of two sections-specifications and methods of testing. Under specification, filament form of saffron based on extraneous matter and floral waste divided into three grades. The grades 1, 2 and 3 should not contain more than 0.5, 3 and 5% of floral waste and 0.1, 0.5 and 1.0% of extraneous material respectively. Chemical characteristics of three groups of saffron as filament or powder in this standard are explained in Table 10-20 (34).

In the second part of French standard, besides explaining methods of chemical and physical analysis, there are methods for

Table 10-20 *Chemical characteristics of saffron in form of filament and powder, based on French standard*

Characteristics	Form of saffron	Acceptable Content		
		Grade-1	Grade-1	Grade-1
Moisture & volatile content	Filament	12	12	12
	Powder	10	10	10
Total ash	Maximum	8	8	8
Acid insoluble ash	Maximum	1	1	1.5
Solubility in water %	Maximum	65	65	65
Picrocrocin	Minimum	70	55	40
Safranal (absorbance in 530 nm)	Minimum	20	20	20
	Maximum	50	50	50
Coloring strength	Minimum	190	150	100
Total nitrogen %	Maximum	3	3	3

determining artificial colors, which might be added to saffron by adulterators. This can also be explained by using column chromatograph and HPLC (35).

10-4-4 Adulteration in Saffron and Methods of Evaluation

Saffron being an expensive spice is liable to frequent adulteration at the hands of dishonest traders for economic gains. Adulteration of saffron has been known since ancient times. In Spain, when Venice was the centre of saffron trade, there was a special police force whose responsibility was to inspect and give assurance about the purity of saffron.

The main adulterants and substitutes of saffron include:
- Coloring the style of saffron with color extracted from stigma or other natural or synthetic colors and mixing them with saffron. This type of adulteration is the most common method of cheating in saffron.
- Adding materials with plant origin to saffron such as petals of sunflower (*Carthamus tinctorius*), petals of *Calendula officinalis*, which is colored with methyl orange, corn silk (stigmas of maize), recoloring of saffron after using it once, adding stigmas of other species of saffron genus, which are shorter and free of pigments like *Crocus vernus* L., *Crocus especiosus* L., *Punica granutum* L., Crushed red pepper, crushed turmeric rhizomes and fibrous roots of various grasses, slender roots of willow, sliced petals of meadow poppy and so on.
- Increasing the weight of saffron by storing in moist stores or spraying water mist over the saffron, treating saffron with oils, honey, glycerin or potassium nitrate, ammonium nitrate, prox, glucose invert sugar, sodium sulphate, etc.
- Increasing fibers of shredded meat colored with saffron water.
- Adding synthetic colors to saffron in order to increase its coloring strength.

• One of the most common ways of saffron adulteration is repeatedly hitting saffron bunches against some hard surface. By doing this stigma breaks down and this stigma is sold as the top quality saffron while the remaining bunch with shorter stigma than the normal saffron is sold as standard saffron.

Methods to recognize saffron adulteration

There are some tests that are based on chromatical reaction. The summary of these methods is presented in Table 10-21 and 10-22. The chromatography methods of saffron adulteration are shown in Table 10-23.

Methods of testing mentioned in saffron standards can differentiate pure saffron from adulterated saffron. For checking the

Table 10-21   *Coloring reaction for detecting saffron adulteration (10)*

Type of material	Indicator	Color
Pure saffron	Sulphuric acid. Aluminium oxide. Hot fehing solution.	Blue, Velvet blue or red hue. Dark orange Brick red
	Carr-Price reaction Adding small amount of saffron to 0.1 g Di-phenyl amine in 20-cc of sulphuric acid and 4 cc of water.	Positive Blue quickly changes to reddish brown but in the presence of nitrate it is not so.
Sandal wood or Brazili	Ether doitrol	Lemon yellow
Poppy	Amonium nitric acid	Greeny gray
Infusion of poppy	Nitric acid	Bright red
Anato	Cs_2, $C_2H_2Cl_4$, $CHCl_3$, $C_6H_5CH_3$, C_6H_6	Bluish red
Synthetic colors	Organic solutants	Pink or red
Glycerol	Denige test	Bluish gray
Sugar	Denige test	Red / Velvet red

Table 10-22 *Reaction of saffron and other natural colored materials to indicators (10)*

Indicator	Saffron	Anato	Carotene, Xanthophylle	Sandal Wood	Turmeric
Dence HCL	Little or no change	Remain orange	No change or a little color reduction	More reddish	Orange-red by adding dense HCL
10% NaOH	Remains yellow	-	No or little change	Velvet to bluish velvet	Orange-brown
Hyposolfit sodium	Low effect	Low effect	Low effect	Lost color and by oxidation changes color	Low effect
Chlorurferric 5% Solution	No change. Sometimes more brownish	No change. Sometimes more brownish	-	Dark velvet or black brown	No change. Sometimes more brownish
With 10%	Low change	-	-	Red (Change is very slow)	Little change
5% Uranil acetate solution	No effect	-	-	Velvet. Lost its color quickly	No change. Sometimes more brownish
Sulphuric acid on dry sample	Blue	Blue	Blue (reaction is very slow)	Red changes to brown	Red

Table 10-23   *Chromatographical methods for detecting adulteration in saffron*

Method	Saffron or adulterant	Position & color
Paper chromatography with bothanol, acetic acid and water (1:1:4) solutant and use of Tantur for pointing	Saffron	Two bright points with Rf = 0.31 and 0.3 in light
-Do-	Turmeric	Yellow, greeny and yellow spots with Rf = 0.98, 0.90 & 0.81
	Sunflower	Blue spot in Rf = 0.98 & 0.95 & green in Rf = 0.53 & 0.23 & red in the beginning
-Do-	*Calendula officinalis*	All spots are blue & green
Thin layer chromatography with Benzidin/Banzan (29/21) solutants	Saffron	No movement in spots Spots with Rf=0.22 and 0.61
	Other materials	in 265 nm and one spot with Rf=0.35 by adding sulphuric acid
Thin layer chromatography with alcohol isobotilic, water, acetic acid (10:5:25) and using alcoholic extract 70%	Saffron	Three visible spots. No Rf given
Paper chromatography with pH enol and water (20:80 w/w)	Saffron	Rf = 0.70
	Sunflower	Rf = 0.35
Solutant amonial 0.88 and water (99:1 v/v)	Saffron	Rf = 0.09
	Sunflower	Rf = 0.66
Solutant sodium chloride 2.5% in water	Saffron	Rf = 0.07
	Sunflower	Rf = 0.42
Solutant 5 cc Ammonia 88% and 95 cc water adding 2 g citrate three sodium	Saffron	Rf = 0.08
	Sunflower	Rf = 0.68

Microscopic methods for detecting saffron adulteration

synthetic colors in saffron thin layer chromatography and HPLC can recognize these colors.

One of the suitable and most precise methods for checking the purity of saffron is the study of anatomical properties of saffron using a microscope. Figure 10-11 shows the transverse section of stigma of saffron.

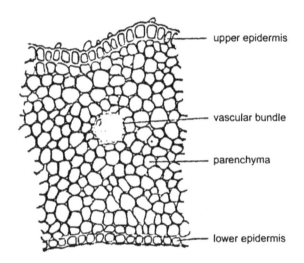

Fig. 10-11   Transverse section of stigma of saffron (18)

This section has the following parts

- A parenchyma formed of polygonal cells or cells with rounded corners with thin walls
- Vascular bundles of round cross section
- An epidermis composed of a row of slightly elongated plate cells perpendicular to the surface of the stigma and covered by a thin cuticle. Some epidermic cells have a small papilla in the middle of their outside wall.

The essential microscopic features characterizing saffron powder are as follows:

- Fragments of the top extreme of the stigma with large, hair like elongated papillas capable of reaching a length of 150 µm (Fig. 10-12).

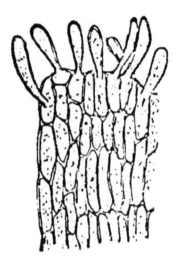

Fig. 10-12   Upper extremity of saffron stigma

- Epidermic deloric of stigmas with small round pupillas (Fig. 10-13).
- Round pollen grains of large diameter (85–100 μm) with a thick, smooth cell wall and a finely granular exine (Fig. 10-14)

In addition parachymatous debris, debris of thin vascular bundles, debris of the epidermis of the style can also be observed (Fig. 10-15) (32, 12).

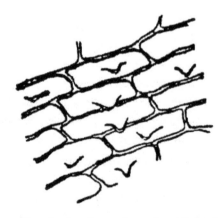

Fig. 10-13   Upper epidermis of saffron stigma (frontal view)

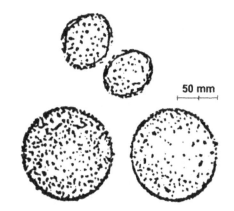

Fig. 10-14 Pollen grain of saffron (magnification x 300)

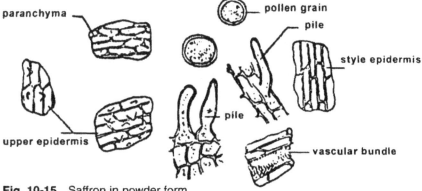

Fig. 10-15 Saffron in powder form

10-5 SUMMARY

The dried stigmas of saffron plant are more than general compounds such as water, carbohydrates, proteins, oils, salts and vitamins and have special compounds. The size and type of these compounds are the principle criterion for evaluating the quality and grade of saffron. The most important pigment of saffron is crocetin (ester di-gentibiose crocetin), which is solvable in water and is one of the best coloring agents in food industry. The main factor in saffron flavour is a colorless glycoside namely picrocrocin. This compound during drying and storage of saffron, changes to safranal. Safranal, alongwith nine other compounds is responsible for aroma and smell of saffron. The quantity of coloring strength, flavour, aroma and smell of saffron is now measurable using spectrophotometry or more precisely by using HPLC.

References

1. Alonso, G.L., R. Varon, R. Gomez, F. Navarro and M.R. Salinas. 1990. Auto-oxidation in saffron at 40°C and 75% relative humidity. J. Food Sci. 55: 595-596.

2. Alonso, G.L., M. R. Salinas and J. Garijo. 1998. Method to determine authenticity of aroma of saffron (*Crocus Sativus* l.). J. Food Protec. 61: (11): 1525-1530.

3. Ashraf Jahani, A. 1993. Historical, commercial and scientific study of saffron. Standard & Industrial Research Organization of Iran. (Persian)

4. Atefi, M. 1999. Effects of freeze drying on qualitative parameters of saffron- A thesis submitted for M.Sc. in Food Science & Technology. Iranian Food Research and Industry Institute. (Persian)

5. Baker, D., M. Negbi. 1983. Uses of saffron Eco. Bol., 37(2):228-236.

6. British Standard 4585. 1994. Method of test for spices & condiments. Part 17: test method specific to saffron.

7. British Standard, 145. 1994. Specification for saffron.

8. Farrell, K.T. 1985. Spices, condiments and seasonings, AVI Press. 197pp.

9. Ghasemi, T. 2001. Saffron, the red gold of Iran. Cultural Society of Nashr Ayandagan. (Persian)

10. Habibi, M.B. and A. Bagheri. 1989. Saffron-agronomy, processing, chemical composition and its standards. Scientific and Industrial Research Organization of Iran, Khorasan Institute. (Persian)

11. Hemmati Khakhki and S. Kamel Rahimi. 1994. Searching & extracting antocianin from saffron petals and study of its stability in a model drink. Scientific & Industrial Research Organization and Iranian Nutrition & Food Industry Research Institute. (Persian)

12. Hemmati Khakhki, A., 2001. Effects of differing methods of drying in preserving saffron quality. Research & Construction in Natural Resources. Vol. 14. No. 2. (Persian)

13. Hemmati Khakhki, A., 2001. Improvement of effective factors on edible color from saffron petal. Agricultural Sciences & Industry. 15: 13-21. (Persian)

14. Hemmati Khakhki, A., F. Majid, N. Khalili, H.R. Zolfogharin and M. Mokhtari. 1997. Processing & packaging of saffron. Scientific & Industrial Research Organization. (Persian) of Iran, Khorasan Institute.

15. Himeno, H. and S. Konosvke, 1987. Synthesis of crocin, picrocrocin and safranal by saffron stigma like structures proliferated in vitro. Agr. Biol. Chem., 512: 2395-2400.

16. Ibora, J.L., M.R. Canovas, M. Canovas and A. Manjon. 1992. TLC preparative purification of picrocrocin, HTCC and crocin from saffron. J, Food Sci. 57(3) 714-716.

17. International Organization for Standardization (ISO) 3632-1. 1993 (E). Saffron (*Crocus sativus* Linnaeus)- Specification. Geneva, Switzerland.

18. International Organization for Standardization (ISO) 3632-2. 1993 (E). Saffron (*Crocus sativus* Linnaeus)- Test methods. Geneva, Switzerland.

19. Institute of Standard & Industrial Research Organization of Iran. 1993. Characterization of saffron. National Iranian Standard No. 259-1. (Persian)

20. Institute of Standard & Industrial Research Organization of Iran. 1993. Methods of sampling in saffron. National Iranian Standard No. 3659. (Persian)

21. Institute of Standard & Industrial Research Organization of Iran. 1997. Saffron-Methods of Testing. National Iranian Standard No. 259-2. (Persian)

22. Institute of Standard & Industrial Research Organization of Iran. 2001. Regulation for harvesting & processing of saffron before Packing. National Iranian Standard No. 5230. (Persian)

23. Institute of Standard & Industrial Research Organization of Iran. 2002. Microbial Specification of saffron & their Testing. National Iranian Standard No. 5689. (Persian)

24. Institute of Standard & Industrial Research Organization of Iran. 2003. Regulation for Establishment of HACCP from Harvesting to Packing. National Iranian Standard No. 6762. (Persian)

25. Institute of Standard & Industrial Research Organization of Iran. 2003. Regulation for Instruments & Equipment Necessary for saffron Packaging Workshop. (Persian)

26. Kamel Rahimi, S. 1993. Effects of temperature and moisture on duration of storage of saffron. Scientific & Industrial Research Organization of Iran, Khorasan Institute (Persian).

27. Khorasan Jahad-Keshavarzi Organization. 2001. Directorate of Statistics & Information.

28. Mannino, S. and G. Ameiooti. 1997. Determination of the optimum humidity for storage of saffron. Rivista-dell-Societa-Italiana-discienza-dell Ali-menta zone, 6(2): 95-98.

29. Molafilabi, A. 1994. Study of the flower components of saffron. Proceedings of Second National Congress on Saffron & Medicinal Plants. Scientific & Industrial Research organization of Iran, Khorasan Institute (Persian).

30. Morimoto, S., Y. Umezaki, V. Shoyama, H. Satio, K. Nishi and N. Irino. 1994. Post harvest degradation of carotenoid glucose ester in saffron. Plant med. 60: 438-440.

31. Nair, S.C., B. Pannikar and K.R. Panikkar. 1991. Antitumour activity of saffron (*Crocus sativus*). Nutrition and Cancer. 16(1): 67-72.

32. Negbi, M. 1999. Saffron (*Crocus sativus* L.). Harwood Academic Publishers.

33. Noorbaksh, R., A. Hemmati Khakhki, R. Razaghi and M. Saberi. 2001. Effects of antinicrobial of saffron stigma and application of different packaging materials in saffron characteristics. Iranian Standard & Industrial Research Organization. (Persian)

34. Norme Francaise (NF) V. 32-120-1. 2000. Safran (*Crocus sativus* L.) Partie 1: Specifications. Association Fraincaise de Normalisation (AFNOR).

35. Norme Francaise (NF) V. 32-120-2. 2000. Safran (*Crocus sativus* L.) Partie 2: Methodes d'essai. Association Fraincaise de Normalisation (AFNOR).

36. Orfanou, O. and M. Tsimidou. 1995. Evaluation of coloring strength of saffron spice by UV-vis. Spectrometry. Food Chem. 57(3): 463-469.

37. Pfander, H. and M. Rychener. 1982. Seperation of crocetin glycosyle esters by HPLC. J. Chromatog. 234: 443-447.

38. Raina, B.L., S.G. Agarwal, A.K. Bhatia and S.G. Gaur. 1996. Change in pigments and volatiles of saffron (*Crocus sativus* L.) during processing and storage, J. Sci. Food Agric.71:27-32.

39. Rashed Mohassel, M.H., A. Bagheri, M. Sadeghi Tehrani and A. Hemmati Khakhki. 1989. Report of a delegation journey to study spanish saffron production. Scientific and Industrial Research Organization of Iran, Khorasan iInstitute. (Persian)

40. Sujata, V., G. Ravishankar and L.V. Venkataraman. 1992. Methods for the analysis of saffron metabolites, crocin, crocetins, picrocrocin and safranal for the determination of the quality of the spice using thin-layer chromatography, high performance liquid chromatography and gas chromatography. J. Chromatogr. 624: 497-502.

41. Torabi, M. 1990. Methods of saffron standard testing in Spain. Scientific and Industrial Research organization of Iran, Khorasan Institute. (Persian)

42. Trantillis, P.A., G. Tsoupras and M. polissiou. 1995. Determination of saffron components in curd plant extract using HPLC-UV-visible pH otodiode-array detection-mass spectrometry. J. of Chromatogr. 669: 107-118.

43. Tsimidou, M. and E. Tsatsaroni. 1993. Stability of saffron pigments in aqueous exteacts. J. of Food Sci. 58(5): 1073-1075.

44. Unknown. 2001. Effects of time of harvesting and processing on microbial population and quality of saffron. Department of Education & Research of Jahad-e-Keshavarzi. (Persian)

45. Ursat, J. 1997. Le safran DV Gatinais. Connaissance et memoires Europeennes.

46. Valizadeh, R. 1998. Use of saffron leaves as a source of fodder. Scientific & Industrial Research Organization of Iran, Khorasan institute. (Persian)

47. Zarghami, N.S. and D.E. Heinz. 1971. Monoterpene aldehydes & isophorone related compounds of saffron. ph ytochemistry. 10: 2455-2761.

CHAPTER **11**

Research Strategies

M. Kafi
College of Agriculture, Ferdowsi University of Mashhad, Iran

11-1 INTRODUCTION

During the 20th century, particularly in the second half, agricultural products have increased remarkably to meet the requirements of the growing population of the world, which has reached the point of explosion. To meet the requirements sufficiently, some of the crops have gone into oblivion. The increase of a product is possible by two means. The first is to increase the area, which is under cultivation and the other is to increase the production in a unit-area (7). Although both these factors have been used in the last century to increase production, the share of yield increment has operated more than the share of increasing the area under cultivation. Take the example of India. In the year 1960 the production of wheat was about 400 kg per hectare but in the year 2000 each hectare of land produced 2400 kg (6). This means that the growth in yield has increased six fold during the last forty years. Not only in India, but also at the world level during the second half of the 20th century the area under cultivation of wheat has increased from 180 million hectares to 230 million hectares. This means that there has been an increase of 30% in areas under wheat cultivation, where as in the same duration the operation of the growth of the wheat production has increased by 150% in every unit area, which has reached from 1000 to 2500 kgs per hectare. The production of wheat also in this duration has increased from 200 million tones to 600 million tones (6).

The same principle also applies on all the field products, where share of yield has increased several times in every unit area as compared to that of the area under cultivation. The important factors that have caused an increase in the yield can be described briefly as given below:

1. The increase of the quotient in the deduction of the produce. This quotient in wheat has reached about 3% to 5%. The same principle also works in other products under cultivation
2. The genetic improvement of the figures related to cultivation and the increase in the potential yield of the cultivation figures
3. The increase in the capability to meet the requirements in the cultivation work. These are the fertilizer stability in that particular area.

During the last thirty years, i.e. from 1971 to 2001, the land for cultivating saffron in Iran has increased from about 3000 hectares to 47,207 hectares, which shows that the increase has been fifteen times. During this span of time the production of saffron has also increased from 17 tonnes to 160 tonnes (6). The result, which is derived by comparing these two figures, is that this increase has taken place by increasing the area of cultivation. During the last thirty years the operation in a particular unit area not only not increased, it has reduced. This situation, perhaps might not take place in any other crop plant. Now, here rises the question, whether there is also further possibility to increase in area of land for cultivating saffron? If the reply is positive, then at the cost of which item can this be achieved? Is the way to increase the yield of saffron blocked?

Taking into consideration that in the regions where saffron is grown, all the sources of water, which can be exploited are being used. Almost in all the fields, where this crop is being cultivated, the negative balance of water also exists (8). This is on account of the possibility that to increase the area of land for its cultivation is at the expense of losing area for other crops. The only other way it to increase the yield per unit area of land, which is under cultivation. This is possible when the plantation of other crops is determined and the land set aside to grow only saffron. This would lead to reduced production of the other crops and cause an increase in the cost of saffron production. One of the foremost examples is Turbat-e-Hayderya district, where the land has been specified for growing saffron by omitting to grow the other crops. The result of this has been that in the last ten years the area where saffron is grown has increased from insignificant number to 15,725 hectares (6).

Here, it is also necessary to point out that the average production of saffron in the specific unit areas in countries like Spain and Italy is two to three times more than Iran and India. Therefore, there is a possibility to produce more saffron in the country by operating more practically. Research has already been done to show increase in production per unit area by adopting a certain method. But there are still some points that have remained vague in the production of this valuable plant. The hard working farmers put in a lot of effort in both cold weather and the burning heat of the sun. Now, it is the duty of scientific research centres to bring under the limelight the obscure dimensions and hand them over to the executive promotion boards so that they may direct those who produce the crop. In this way, it may be that their support, the people who are engaged in this business may make Iran the leader as it was in the past (9).

11-2   RESEARCH PROCESS OF SAFFRON

Saffron is a valuable plant and has great economic importance; however, not much research has been done on saffron as has been the case with other plants. In Agricola site (This is an important site for agricultural products and publishes several articles on agricultural topics) it has been pointed out that during the years 1969 to 1996, only 91 articles have been written on saffron, which is approximately three articles per year. This has been pointed out in the Fig. 11-1, which indicates the distribution of the articles in different years. It is very clear from this data that not much research work has been carried out on saffron. This subject was written between 1975 and1979 and again 1983–1984, but after that there has not been much written about this subject. Attention on this subject had been paid on average during the years from 1975 to 1979 and from 1983 to 1984, while after that, on average, there has been written less on this subject (11). The maximum number of articles in any one year has been only ten. It has to be pointed out here that the articles that have been written also include those which are related to saffron's chemical composition and its medicinal properties.

During the last two decades the research that has been done on a large scale by the Scientific and Agricultural Research Organization of Khorasan has mostly concentrated on the agronomic part of saffron, which has pointed out its many unknown agricultural values and properties (see Table 11-1). But, in spite of all this, much more work has to be done to solve the problems faced in developing the production of this plant and to learn accurately qualities related to its ecology, physiology, morphology and cytology. Here, in this section, a brief account has been given on the research work dealing with saffron:

11-3 POLICIES AND STRATEGIES FOR RESEARCH ON SAFFRON

It is possible that under the available resources, and based on the produce of a particular product, research priorities may change. However, the sub-headings that have been given below are on subjects that will make clear the way for research in different dimensions to encourage production and cultivation of saffron.

1. The agronomical studies

At present, saffron corm is planted either in the form of piles or separately, which is decorated with different forms and designs. The

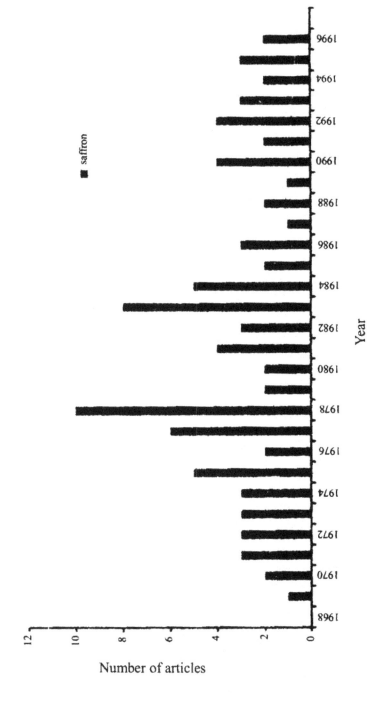

Fig. 11-1 Distribution of articles published on saffron based on the statistics recorded by the Agricola Information Bank.

Table 11-1   *A list of saffron research projects that have been completed by the Scientific and Industrial Research Organizations of Iran, Khorasan Institute.*

S.no.	Title of the project	Name of the researcher
1.	Optimization of the factors by which the color can be produced from the petals of saffron to dress food.	Abbas Hemmati
2.	The survey of the problems related to saffron.	Abbas Hemmati
3.	The study of biochemical properties of saffron in the South of Khorasan	Gholam Ali Kamali
4.	Evaluating the physical and chemical properties of the water and soil related to the processing of saffron.	Hamid Shahandeh
5.	The effect of the saffron corm in the flowering of saffron	Bahzad Sadeghi
6.	The exploitation of the weed in the fields of South of Khorasan	Mohammad Hassan Rashed mahassel
7.	The preliminary survey of the diseases of saffron	Bahrooz Jafarpour
8.	The effect of summer watering to increase the production of saffron	Bahzad Sadeghi
9.	The effect of storage, the date of planting the corm for the flowering of saffron	Bahzad Sadeghi
10.	The effect of chemical fertilizers and manure on producing the leaves, corm and flavour of saffron	Bahzad Sadeghi
11.	The effect of leaves nourishing to increase the production of saffron	Mohammad Hosseini
12.	The study of nourishing the animals on the leaves of saffron	Reza Valizadeh
13.	The search and extract of anthocyanine of the petals of saffron and examining of the durability of that in a model drink	Abbas Hemmati
14.	The first phase of the semi-industrial plant to produce the colour (*anthyocyanine*) to dress the food with the petals of saffron	
15.	The examination of the heat and moisture for the protection and preservation of saffron	Rahimi and Hemmati
16.	Examining the effect of time of plucking flowers to maintain the quality and quantity of saffron.	Abbas Hemmati
17.	Examining of the effect of different methods of drying and preserve the quality of saffron	Abbas Hemmati

(Table 11-1 Contd.)

(Table 11-1 Contd.)

18.	The processing and packing of saffron	Rahimi & Hemmati
19.	The search and extract of anthocyanine of the petals of saffron and examining of that in a model drink	Rahimi & Hemmati
20.	The study and evaluation of the social and economic effects of the ten-year research on saffron	Hosseini & Hemmati

month recommended for the cultivation of this plant is the month of Khordad (May-June) on the basis of the results that have been derived after research in this field (10). The corm blossoms in four months time by the beginning of the month of Mehr (September-October). The depth at which the corm is planted differs from place to place and any particular alteration in the systemhas not been taken into consideration. It is essential that the research work that has been accomplished and is ongoing, should be given to the producers for the promotion of this plant (10).

2. The examination of the causes by which saffron cannot be planted a second time in a particular soil

Farmers who are engaged in the business of producing saffron in Iran and also in other countries believe that the corm once removed from the soil cannot be planted again in the same soil. But, still a scientific reason to support this belief has not been given. In the physico-chemical and mineralogical examination and full analysis of the soil this has not yet come into notice, which soil is more useful than the other and what is the difference between the soils, which are under cultivation of saffron (3, 4). Some of them believe, as there is aliphatic substance in the sub-soil, therefore, the corm cannot grow there if planted the second time. This subject needs more research on a larger level, which has several aspects like pedology, plant physiology and the others.

3. The cultivation of saffron with other plants

Saffron is a plant which has a composite structure. This plant, whether from the point of view of space or time, can live for upto a year without the need of any space or light. In the same manner, the structure of its roots and the corm is such at a particular depth that is absorbs water and alimentary elements. Therefore, the cultivation of saffron is done with other plants with the purpose to benefit more from the water and soil. Though much research work has not been done in this field, but the producers should be made aware of the advantages and disadvantages of cultivating saffron with other plants.

4. Examining the samples of corms to be sown

One of the most important discussions of physiology of plant life is maintaining a balance between the matter nurtured in the cistern and the capacity of the preserved matter kept in the deposit. In saffron, leaves serve as the source and the corms function as the sink. But, in the beginning, the daughter corm turns into source from deposit being the mother corm. In particular, some part of the deposited mother for the daughter corm, there are several questions that rise about the share of the mother corm in the formation of little corms, the share of photosynthesis functioning to get the corm filled and finding the limitations of the source or deposit (8, 9, 10). These questions should be answered by doing further research on the subject.

5. Examining the beginning and evolving of saffron corm

In Iran exhaustive study has not been done so far on the beginning and the formation of the corm of saffron. Whatever information we have at present is based on the articles that have been written by authors from Kashmir and Azerbaijan. The articles that have been published provide the required information but there is still the need to learn about regional climatic and agricultural effects, time taken to produce the crop in a particular field and the different stages of growing the corm. It is necessary that the universities and research centers do exhaustive studies on the subject.

6. Examining the needs of temperature for the saffron plant

One of the complications of saffron is the requirement of heat for this plant. From this point of view the plant reacts quite contrary to other plants. This is in the sense that it grows with the start of the winter season and the growth stops when summer sets in. In the same manner, it is possible that in two different regions, though there may not be parametric difference of temperature, in meteorology, but there are different reactions. Even the most preliminary reports have not so far related the base temperature required for growing for each stage of its growth, the required temperature, photosynthesis and the maximum and minimum tolerable temperatures. The figures that have been produced are mainly based on the growth of this plant in particular regional condition.

7. The study of the reaction of saffron at the intensity of heat and the quality of light

Scholars have not studied the subject concerning the availability of light in the saffron growing regions. They have also not paid attention to the intensity of the quality of light but the fact is that there are long cloudy days during the growing season of saffron. Covering by snow during some part of the winter season over the saffron plant, casting of shadows by the leaves over each other and also by the weeds over the flowers of saffron, the intensity and quality of light received through the leaves as compared to the natural sunlight is always changing. There is also a possibility that some fluctuations that occur at the time of growth and production of saffron may be due to the condition of light that dominates during the growing season is hoped that the research done in future will bring more light to the subject.

8. The equilibrium of photosynthesis and measurement by related parameters

The equilibrium of the growth of the plant is a product of the level of photo synthesizer in the equilibrium of photosynthesis. In case of saffron the quotient of the level of the leaves has been studied, but there does not exist any report based on the equilibrium of photosynthesis on the unit of the level of the leaves. In the same manner the velocity of the ostiole of the plant and the equilibrium of different kinds of chlorophyll that exists in the leaves particularly with the attention of the colour of the leaves has not been studied. When research is done on this subject, some important physiologic quotients will be determined.

9. The study of the reaction of the plant to biotic stresses

Much research work has been conducted on the pests, weeds and plant diseases in connection with living tensions. As the greater part of growing of this plant takes place in the winter season, it, therefore, helps to control the pests and some of the diseases that are caused by the weeds. In spite of all the different reports that have been given about the loss and damage that is brought to saffron by pests and weeds, there is still more need to research this subject.

10. The study of the reaction of the plant to abiotic stresses

Saffron is a plant, which under the condition of the environmental growth during the time of its growing comes across several different environmental tensions. The most important environmental tension, which are being faced by the plant, are the tension of dryness, the tension of cold and the tension of salinity. Subsequently, in the heaps of local saffron there is reaction of all these tensions that appear in the form of its varieties. The genotypes, which give stability to it, will cause evolution in the production of saffron. Though studies have been made in the field of the reaction of the plant that brings tension to it due to the scarcity of water, but there are still several other unknown dimensions of the reaction that cause tension due to scarcity of water. Besides that there are several other tensions that have not been reported so far.

11. The different usage of the other parts of the flower and the small corms

On average, 80 kg of saffron flowers produce one kg of stigma and the separate style while as much as 75 kg is the wastage that is removed from it. The tests that have been made (4, 5) on the flower show that the flower contains several other effective compounds that can be used in several other industries. In the same manner, the small corms are not useful for replanting and, thus, these cannot be used as seeds to grow saffron, but these still contain useful compounds, the nutrients of different kinds of carbohydrates, lipids and proteins. With research being done in this field, proper ways can be found to make use of the flower and better use of the corms for providing new resources of income for the growers.

12. Changes in the structure of canopy and an ideal designing for saffron

Scholars who have specialized in the physiology of agricultural plants have commended one type of plant. Experts on reforming plants follow the plan so that they may succeed in attaining their object (7). Notable problems, which are related to saffron and should be pointed out, are:

   i. The measurement of the leaves, particularly their length
   ii. The colour of the flowers
  iii. The angle of the leaves of saffron
  iv. The number of leaves suitable to meet the nutritious needs of the corm
   v. Productive capability of the corm
  vi. The required length of the corm
 vii. The proportion between the stigma and the rest part of the body of the flower
viii. The length of the style

13. The reform of the figures of saffron

Among agricultural crops like wheat, about 50% increase in the yield in the past half-century is indebted to genetic improvement. These genetic changes have taken place by means of selecting the proper genotypes, the hybridization of the lines with required qualities with conventional cultivars and importing high yielding cultivars from all over the world. In case of saffron, there is no report based on the breeding a cultivar or the genetic variation in Saffron that is cultivated in different parts of the world (there has been only one instance that has been published). The first step that should be taken in this direction is that study different varieties that exist in the heaps of cultivable saffron in the different parts of the world and start breeding programs in order to introduce reliable cultivar (s) to the farmers.

14. The creation of the genetic variation

Variation is the base of stability and the result of the crossing between the different genotypes. Generally the agent that produces variation is the crossing between the different varieties of one species or different species naturally or artificially. When saffron is sterile and the seed does not take form, its proliferation is done by asexual means. Therefore, there is not any classical method of crossing that may take place by the breeding of the plants. Perhaps, the appropriate means of breeding can be the new ways of biotechnology. The search has been made to apply the radiation of gamma rays on the bulb of saffron, which resulted in the improvement of the yield of the plant (1). This field can be a subject of interest for research in the countries where saffron is a source of income for their farmers.

15. Stability in the yield of saffron

Though all efforts are being made by the saffron-producing farmers to get the maximum results for their efforts, but in spite of it, it has been found that the yield varies from year to year. There are many horticultural trees that show fluctuation in fruit production year by year. They usually produce high yield in one year and no or low yield in second year. This might be because they have enough reserves of photosynthetic materials from the last year and that gives them this capability to produce plenty of fruits. But in the next year they lose the essential deposits to start producing fruit and during the growing season, the tree or plant compensates its lost deposits. In the case of saffron too the scale of rain and the technical management of the farm might be effective. Research should be done in this regard.

16. Use of biotechnology and molecular genetics for breeding of saffron

Since saffron is a sterile plant, which proliferates by means of a bulb, this plant does not get crossed with other ones as usually happens with other plants. On the other hand, biotechnology, molecular genetics and tissue culture have made remarkable progress recently, and they are able to make easy and shorten the long classical breeding process. It is, therefore, these technologies that might solve the problems, which are in the way of saffron breeding.

17. Ecological demarcation of the areas for the growth of saffron

In the past saffron was planted in vast areas in different parts of Iran. But during the last century, the South of Khorasan, some parts of Fars, Yazd and Kerman provinces have become important centres for producing saffron. But to renew the promotion of it in Iran on the basis of ecological demarcation, minute study should be made from the ecological, social and economic point of view. It seems possible that if there is area wise demarcation for the development of saffron, the systematic plans can open new avenues for those who are interested in the growth of saffron.

18. Packing, exploring markets and export of saffron

For the briskness of the production of saffron there is a need for strong backing in areas like modern packing, marketing and transporting to the markets. If these measures are adopted, Iran can remain the leader in saffron production. It is hoped that parallel to the studies related to the growth of agriculture and breeding, research will continue in post-harvest of saffron.

19. Uptodate economic studies

Contrary to the results that have been derived from the research done on the basis of the short duration of the life of the fields where saffron is cultivated and harvesting of saffron being useful only for five years from the economic point of view, the producers of saffron in South Khorasan have preserved it to keep saffron in the same land even for fifteen years. To reach the root causes of this matter, there is need of further research, particularly from its economic point of view. So this can be explained to the farmers what loss they are suffering per year in the growth of the life of the field. This work should be done on regional level. In the same manner, the five-year life span of the field is also connected with different factors like the management of the field and the fertility of the soil. There should be some correlation between these factors and the life of the field.

20. The establishment of the research center for saffron

The above mentioned nineteen clauses and perhaps several other untold problems that have not been solved so far related to the cultivation of saffron and its processing on the basis of scientific and economic means require the need to establish an independent research centre so that the different agricultural, breeding, agro-technical, processing, marketing, packing problems and extracting of essential components by scientific methods can be examined and the results derived after the examination should be handed over to the producers and processors who are engaged in the business of this valuable product. Such a center should be established at such a place where saffron is grown. Obviously, the funding for establishment of such a research center could be compensated by growing saffron on scientific basis.

4-11 SAFFRON ON INTERNET

At present, besides the scientific journals and books, there are also websites that provide more information on this subject. Internet with all the peculiarities that can be specific for an individual is different from the books and journals on this subject, which means that it provides information free of cost. Presently, universities, institutes, publishers, foreign companies and individuals have accepted its need for gathering information and publishing their works. Though there is not much scientific information available, but in spite of that, efforts have been made here to provide some addresses that can be useful in this regard:

http://www.saffron.gr/

This site provides information about the history of saffron produced in Greece, its medicinal properties and its application.

http://www.crop.cri.nz/broadshe/saffron.htm

This site has been designed by Malcom Douglas for the Red Bank Research Centre in New Zealand. This site gives general information on saffron including its ecological needs, agricultural requirements, its production and the position of the market. It should be noted that since saffron is not an agricultural crop in New Zealand, the information is not current.

http://www.holisticonline.com/herbal-med/-herds/h163.htm

This site mentions the medicinal properties of saffron and the quantity in which it should be used. General information on saffron has also been provided.

http://www.chromadex.com/phytosearch/saffron.htm

This site lists the chemical properties found in different parts of this plant, particularly in its stigma.

http://www.edis.ifas.ufl.edu/mv128

This site, prepared by J. M. Stephens at the University of Florida, provides useful information for those cultivators who want to promote the plantation of saffron.

http://www.novinsaffron.com

This site belongs to Novin Saffron Company (NSC) that has been very active during the past few years for packing, processing and exporting saffron. This company has been able to provide information about the standard and scientific methods for packing and displaying saffron. NSC also lays emphasis on the significance of Iranian saffron.

http://www.tarvandsaffron.net

This site belongs to Tarvand Saffron Trade Company. The company has been dealing in saffron for the last thirty years. The site provides general information on saffron.

http://www.iranagrofood.com/Saffrom.htm

This site has been prepared by Sadr Qayeni Saffron Company and provides useful general and agricultural information on saffron.

http://www.erisiansaffron.com

This site provides chiefly commercial and standard information on saffron.

http://www.iransaffron.org

This site belongs to the Centre of Information and Research on Saffron. This is the most complete site that provides information about saffron. The site was devised in order to provide the information on the history, botanical information, chemical compositions, different methods of flower picking, processing and the list of different research projects. Presently, the summary of the output, world trade of saffron and other information has been inserted in the site. What is more important is that this site is available in English, Persian and Arabic. This valuable base has been designed by a non-governmental body, which sought technical help from the food industry and the Ministry of Industries.

http://www.spanish-gourmet.com/azafran/Saffron.htm

This site, prepared in Spain, provides information on the history of saffron, its botanical properties, composition and its different uses. In this site it has been shown that Iran is one of the largest saffron producing countries but the best saffron is produced in Spain.

http://www.saffron-golnaz.com/saffron.htm
This site gives statistical figures of saffron, its medicinal properties,
consumption and genuineness have also been recorded.

11-5 SUMMARY

Although, saffron is not a new crop plant and it has in fact a deep-
rooted ancient history in Persia as well as the world, but it has
benefited least from modern science. The remarkable progress that
started in the yield of many agricultural crops during the second half
of the twentieth century has been neither good nor remarkable for
saffron. On the contrary the situation in Iran is quite the opposite
with regard to the production of saffron when compared with past
production. This trend, on one hand weakens the power of
economic competition with the other agricultural crops and on the
other hand the farmers of this plant on the grounds of getting lesser
income from this crop, would not be able to compete with saffron
producers of the world. It is, therefore, imperative that more
emphasis is laid upon the research related to the ecological and
agricultural requirements of saffron. This matter becomes more
serious when the principle findings of the other plants cannot be
applied to saffron. This is because the system of production, the
season of harvesting and process of breeding of the other plants do
not correspond with saffron. It is also necessary to point out here
that presently this is also a possibility that the farmers may not
welcome the product of this plant in terms of the income earned. It is
also possible that the producers, with the purpose of plentiful growth
and providing the means of subsistence for their families, may cut its
production every year. This is in the sense that there is a possibility
that there may be other means of income that may be a substitutes to
planting saffron and the level of saffron cultivation may come down.
It is now necessary that there should be a multi-dimensional
research by establishing a research centre for saffron in one of the
saffron producing regions of Khorasan so that this center may be able
to fill the vacuum in this research. There is no doubt that presently

the Scientific & Industrial Organization of Iran in its branches at Khorasan and Fars provinces has some good plans for saffron research. The pole of the scientific agriculture is the center of excellence in faculty of agriculture, Ferdowsi University of Mashhad, which has given first priority to research work on saffron.

References

1. Akhond-zadeh, I. M. and R.Sh. Mozafarova.1975. Study of the effectiveness of gamma irradiation of the saffron. Radiobiologia 15: 319-322.
2. Gharaee, H.A. and M. Baigi. 1991. Study the changes of physio-chemical and mineralogical characteristics of soils under Saffron cultivation in Stahban region. Scientific & Industrial Research Organisation of Iran, Shiraz Institute.
3. Gharaee, H.A. and A.R. Razaee. 1993. Effects of period of saffron cultivation on soil microelements in Estahban region. Scientific & Industrial Research Organisation of Iran- Shiraz Center.
4. Hemmati Khakhki and S. Kamel Rahimi. 1994. Searching & extracting antocianin from saffron petals and study of its stability in a model drink. Scientific & Industrial Research Organisation and Iranian Nutrition & Food Industry Research Institute.
5. Hemmati Khakhki, A., 2001. Improvement of effective factors on edible colour from saffron petal. Agricultural Sciences & Industry. 15: 13-21.
6. Gharaii, H.A. and A.R. Rezaii. Effect of saffron cultivation on trace elements on soil in Stahban. Scientific and Research Organization of Iran, Shiraz Institute, annual report.
7. Khorasan Jahad-Keshavarzi organization. 2001. Directorate of Statistics & Information.
8. Rahimian, H. and M. Bannayan. 1996. Physiological principles of plant breeding. Mashhad Jahad Daneshgahi Publisher (Persian).
9. Sadeghi, B. 1989. Effects of chemical and animal fertilizers on leaf, corm and yield of saffron. Scientific & Industrial Research Organisation of Iran, Khorasan institute (Persian).
10. Sadeghi, B. 1993. Effects of corm size on flowering of saffron. Scientific & Industrial Research Organisation of Iran, Khorasan Institute (Persian).

11. Sadeghi, B. 1996. Effects of storage and planting date on flowering of Saffron. Scientific & Industrial Research Organisation of Iran, Khorasan Institute (Persian).

12. www.newcrops.uq.edu.au, Crocus sativus.htm.

Index

Printed in the United States
by Baker & Taylor Publisher Services